Public Landscape Integration: Building Landscape

公共景观集成
——建筑景观

金盘地产传媒有限公司　策划
广州市唐艺文化传播有限公司　编著

中国林业出版社
China Forestry Publishing House

序言

随着公共景观设计在世界各城市的复苏以及人们日愈增强的意识,公共景观设计的重要性在政府的大力宣传以及支持下显得尤为突出。通过设计公共景观,城市环境得到提升,城市的身份也得以在公众面前建立,而且城市街道的形象会得到提高。注重功能性,安全性,美学功能的公共景观不仅可以促进一个城市的发展,还可以提高城市居民的生活水平。因此,公共景观设计并非只是针对一个人或是一圈人,而是适用于所有的人。

景观不仅存在公共环境中,而且建筑本身也是景观。随着人们对生活环境以及水平的要求日愈变高,人们更希望看到一些新鲜的独特的景观。本套书除了囊括国内外知名的公共景观设计,还包括各种独特的建筑景观。例如挪威 Ornesvingen 观景台,不仅拥有优美的建筑结构,而且将周围的环境映衬的更加壮丽,吸引了很多来自全世界的游客。

在城市中,街道起着决定性的作用。从当初的交通运输作用,现在的街道具有全面,强大的作用。而人们也希望穿梭于舒适,功能齐全,美观的街道。这样一来,街道设施,如:座椅,自行车停放架,候车亭,照明设施等等都有了新的角色去扮演。

公共景观设计要求公共景观和设计二者和谐统一。同时,公共空间中的设施也需要合理设计与安排。

本套书共有三册,每册书各有重点,特色。形象的图标带来更加直接,丰富的信息。第一册是"公共景观",主要包括国内外知名的广场,公园和游乐场设计。第二册是"建筑景观",主要包括以下三个类别:建筑造型和建筑结构,观景台,标识。创意的建筑造型和建筑结构,总能带给人新鲜的感觉。大胆的观景台设计,不仅吸引着游客,而且值得人们参考设计。极具功能作用的标识设计,形象而直接。第三册是"街道设施",主要包括:座椅,自行车停放架,候车亭,照明设施,自动饮水器,垃圾桶和树木防护装置。这些街道设施不仅覆盖面广,而且具有强大的功能性,更加美化了城市街道。

这套书在一个更高的层次上来诠释良好公共环境的形成,可以提高人们的生活水平。特别是随着城市环境数字化的加强,舒服的公共空间设计和便利的公共设施让人们的生活变得越来越高效和充满活力。

Preface

With the revitalization of public environment design in cities around the world, as also the increased recognition of citizens, the importance of public design has emerged for the construction of a guideline which is maintained and managed by government integratively. Through the improvement for city environment by public design, the establishment of city's own identity and the creation of street image in a view of pedestrian is to get higher. A quality of a city is reflected by functionality, safety and aesthetics simultaneously. Therefore, the meaning of public environment design is not for some unique, specific or a minor group, but for several people to use or see for the creation of beautiful city landscape, which also means a design for all kind of people.

Landscape not only exists in the public environment, and the architecture itself can also be a landscape. With people's increasingly high requirement of living environment and living level, people prefer to see some fresh and unique landscape. In addition to including well-known public landscape designs, this book also includes a variety of unique architectural landscapes. For example, Norway Ornesvingen viewing platform not only has a beautiful architectural structure, but also makes the surrounding environment more spectacular, and attracts a lot of tourists from all over the world.

In a city, the street plays a decisive role. From the start, the street was just used for transportation. However, the street now has a comprehensive and powerful role. At the same time, people want to walk through a comfortable, functional and beautiful street. As a result, street furniture, such as bench, bicycle rack, bus shelter, lighting facilities, etc., have a new role to play.

The public landscape design requires public landscape and design to be of harmony and unity. At the same time, the facilities in the public space need reasonable designs and arrangements.

This book contains three volumes, each with unique focus and feature. The icon images bring more direct and rich information. Volume 1, named "Public Landscape", includes famous squares, parks and playgrounds design from all over the world. Volume 2 is "Architectural Landscape", mainly including the following three categories: Architectural Formative Arts and Structure, Observation Platform, Signage. Innovative architectural formative arts and structure always give people a fresh feeling. The bold observation platform design not only attracts tourists, but also is worth referencing. The signage is functional, image and direct. Volume 3, "Street Furniture", includes Bench, Bicycle Rack, Bus Shelters, Lighting, Drinking Fountain, Trash and Trees Protective Device. These street facilities not only cover a lot, and with powerful features, make the city streets more beautiful.

The purpose of this book is to design a book with applied examples of the amenity elements in the street furniture and to accomplish the public facilities on a higher level as one of the design method to upgrade the quality of citizens' lifestyle in the city environments. Especially with the stress of digital city environments development, comfortable public space and street furniture make citizens' lifestyle more energetic and effective.

目录 Content

第二册 建筑景观
Vol.2 Architecture Landscape

建筑造型和建筑结构
Formative Arts ·· 010

"点"立面平价购物中心
Dot Envelope, Low Cost Shopping

滚动的房子
Roll it Experimental Housing

Infomab10 展览亭
Infomab10 Pavilion

和平大桥
Peace Bridge

香奈儿移动艺术展馆
Chanel Mobile Art Pavilion

雅拉艺术区
Yarra Arts Precinct

停车场自动售票机凉亭
House Parking Ticket Machine

Solearena 休闲椅
Solearena

2011 生态展馆
Eco Pavilion 2011

Aero 展馆
Aero Pavilion

Times Eureka 展亭
Times Eureka Pavilion

南加州建筑学院毕业典礼临时大厅
Graduation Pavilion 2011

伯纳姆展厅	Burnham Pavilion	日落教堂	Sunset Chapel
Hexigloo 展厅	Hexigloo Pavilion	蓝色森林	Blue Forest
Tverrfjellhytta 野生驯鹿观景亭	Tverrfjellhytta	犹太人被驱逐出境纪念馆	Jewish Deportation Memorial
壁炉式儿童游乐场	Fireplace for Children	Crater Lake	Crater Lake
温尼伯湖冰上庇护所	Winnipeg Skating Shelters	马丁路德教堂	Martin Luther Church in Hainburg
"明暗"投影咖啡店	Shading Surface	口红森林	Lipstick Forest
Calder Woodburn 休息区	Calder Woodburn Rest Area	O-STRIP 展厅	O-STRIP Pavilion
迈阿密帐篷	Design Miami/ Tent / Moorhead	ONE Kearny 大厅	ONE Kearny Lobby
顶棚和亭子	A Canopy and a Pavilion	环形穹顶	Ring Dome
聚会的地方	The Meeting Place	银色公园码头办公楼	Silver Park Quay
USCE 购物中心	USCE Shopping Centre	巨大的波浪形木材装置	Gigantic Timber Wave Installation
Grafenegg 展馆	Grafenegg Pavilion	空灵的钟声挂桥	Ethereal Chimes Hung From Bridge
洛尔卡广场	Lorca's Square	城市雕塑	Urban Sculpture
罗马尼亚 ZA11 亭子	ZA11 PAVILION in Romania	本迪戈的中国区	The Bendigo Chinese Precinct
绿色小酒馆室内设计	Green Bistro Interior Design	夏季展厅	Summer Pavilion
电视秀凉亭	Gazebo for TV Show	豪华的帐篷建筑	Regent's Place Pavilion

空间灵魂展馆
Movie by Spirit of Space

大运河广场
Grand Canal Square

投掷
Drop

Hyparform 纵帆
Hyparform Vertical Sails

非线性装展馆
NonLin/Lin Pavilion

水晶网
Crystal Mesh

墨西哥独立200周年纪念"火炬"
Bicentennial Torch

Tubaloon
Tubaloon

3form 新设计产品
3form New Ntunning Products

Casalgrande 陶瓷云
"Cloud" Casalgrande Padana

3013 装置艺术
3013 Installation

奥迪百年雕塑
Audi Centenary Sculpture

CAAC 广场
CAAC / Paredes Pino

Cirkelbroen 环形桥
Cirkelbroen

星云
Nebula / Cecil Balmond

FibreC 展馆
FibreC

教学研究展馆
Research Pavilion

粉红色的球
Pink Balls

高架通道
Walking along This Elevated Pathway

Surface Deep 园囿
Surface Deep

云
The Cloud

纯白 Ecoresin
Pure White Ecoresin

未来之塔
Pylon for the Future Competition

C78 枝形吊灯
C78 of Chandelier

100 个稻草人军队
Army of 100 Scarecrows

红球
Red Ball

一阵狂风
A Gust of Wind

巨大的数字折纸老虎
The Giant Digital Origami Tigers

绿色中空体
Green Void

贝克大街 55 号
55 Baker Street

Mikrocop 数据存储办公楼
Mikrocop Data Storage Office Building

Wykagyl 购物中心
Wykagyl Shopping Center

观景台
Observation Platform······170

黄色树餐厅
Yellow Treehouse Restaurant

蒂罗尔山顶观景台
Top of Tyrol

穆尔河瞭望塔
Observation Tower on the River Mur

Woolwich 观景台
Woolwich Lookout

邦代海滩勃朗特海岸步行道
Bondi to Bronte Coast Walk Extension

博特尼港观景台
Port Botany Lookout

挪威 Ornesvingen 观景台
Ornesvingen

Tungeneset
Tungeneset

Flydalsjuvet 峡谷卫生间
Flydalsjuvet

阿尔塔米拉诺散步道
Paseo Altamirano

观鸟台——挪威旅游线路项目
National Tourist Routes Projects

水袖天桥
Long Sleeve Skywalk

海边景观
Microcostas

Gudbrandsjuvet 观景台
Gudbrandsjuvet

Trollstigplataet 观景台
Trollstigplataet

洛格罗尼奥鸟类观测台
Crnithological Observatory

标识
Signage······208

美国匹兹堡儿童博物馆
Children's Museum of Pittsburgh

Halle F 音乐厅指示牌
Halle F Signage

马里兰艺术学院
Maryland Institute College of Art

新教学大楼
New Academic Building

芝加哥艺术学院
The Art Institute of Chicago

第五大道 623 号
623 Fifth Avenue

多伦多皮尔逊国际机场
Toronto Pearson Airport

高架公园
The Highline

Cogeco 公司总部标识
Cogeco Headquarters Signage

美发屋临街面标识
Hairstyle Interface Signage

埃斯托伊宫酒店指示牌
Palace of Estoi Signage

希拉约翰逊设计中心
Sheila C. Johnson Design Center

长崎县美术馆标识
Nagasaki Prefectural Art Museum Signage

刈田综合病院标识
Katta Civic Polyclinic Signage

大型中庭图形化标识
Storehagen Atrium Graphic Signage

兰斯阿姆斯特朗基金会总部标识
Lance Armstrong Foundation Headquarters

枫忒弗洛皇家修道院标识
Signage for Abbaye Royale Fontevraud

医学院道路标识总体规划
Medical School Wayfinding Master Plan

福尔斯克里特高山度假村
The Falls Creek Alpine Resort

ANZ 中心标识
ANZ Centre Signage

Urban Attitude 礼品店标识
Urban Attitude Signage

成人教育中心标识
Centre for Adult Education Signage

华盛顿大学萨弗里厅标识
Savery Hall, The University of Washington

费尔班克斯国际机场标识系统
Fairbanks International Airport Signage System

华盛顿大学康普顿工会大厦
WSU Compton Union B/D

任天堂公司标识
Nintendo Signage

珠宝世界杂志社标识
Jewel World Signage

盖顶人行通道
Covered Pedestrian Crossing

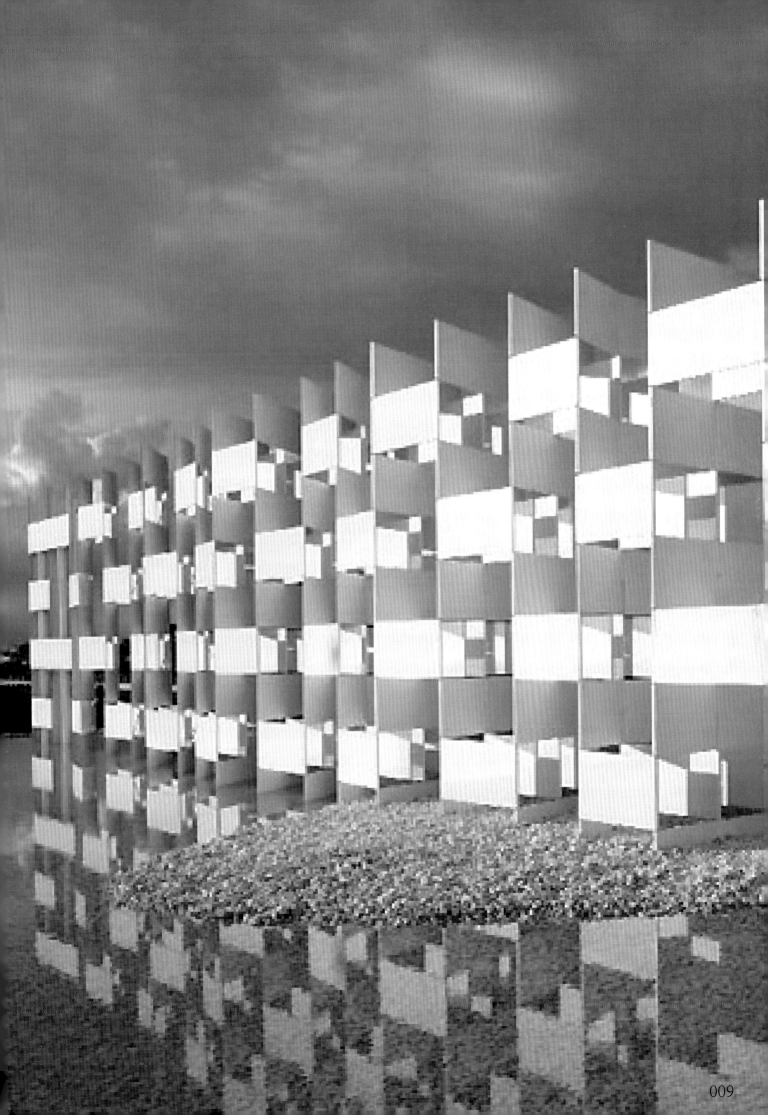

建筑造型和建筑结构
Formative Arts

建筑造型和建筑结构

"点"立面平价购物中心
Dot Envelope, Low Cost Shopping

项目档案

设计：OFIS建筑事务所
项目地点：斯洛文尼亚，卢布尔雅那
面积：2 000平方米

Project Facts

Architect：OFIS arhitekti
Location：Ljubljana, Slovenia
Site Area：2,000 m²

这个项目位于斯洛文尼亚的一个遗址，原先是一个古老的屠场综合体，包括一个水塔和屠宰大厅。

项目旨在重造一个水塔，恢复其原始的风貌，以及利用原有的大厅立面设计一个新的购物中心。客户的要求是建造一个长宽高分别是46米，42米，7米的混凝土大厅，而且室内的设计需与其他连锁店的标准一致。另外，项目的预算非常有限。

新的购物中心的三面都有顾客通道和停车区域。所以这三面所需要的预算只能是一般情况下一面的预算。

整个立面是阶梯式的，这样一来就削弱了整个建筑立方体结构的呆板性。因为有限的预算，设计师们经过精确的计算，决定将整个立面分成三个不同层次的表面结构，20%是由青铜色的金属片构成。而且，金属片上钻有不同大小的圆，这样一来就减少了成本预算。从金属片上钻出的圆也得到了充分利用，安置在另一层混凝土表面上。不同大小的圆形金属片在表面上依次排开，并由绳索相连，很是特别。

Formative Arts

013

建筑造型和建筑结构

Formative Arts

The existing site is listed as an industrial historical area with buildings of an old butchery complex, which included the water-tower and old butcher hall. Demand of National heritage was to rebuild the tower as it was in original and to integrate the main facade portal of old hall in front of the planned new shopping mall. The client permission and expected plan was prefabricated concrete hall of 46x42x7 meters. Interior and all internal finishing is done as "Mercator standard", which is prescribed by shopping company's chain for all their standard malls. Construction Company budget was extremely limited.

The new shopping mall has customer approach and parking facilities on three sides of the building. Therefore it was important to cover three sides with final decorative finishing with the budget of one side only.

The pattern was based on different stepped elevations in order to soften basic cube shell. The designers took and calculated the surface that would be needed for one side. The surface was divided into three elevated surfaces. The surface which could fit into the budget for the facade made use of basic metal sheets which were painted in bronze structured color. After cost evaluation, only 20% of the concrete shell could be covered with the metal sheets. So the sheets were perforated with holes in different sizes. Furthermore the cut metal circles from the sheets were used and arranged on the rest of the facade surface. The use of the holes and circle ornamental patterns were justified with the commercial merchant's company "Mercator pika". For the rest of the budget which remained, the metal chain rope was created around the metal dots.

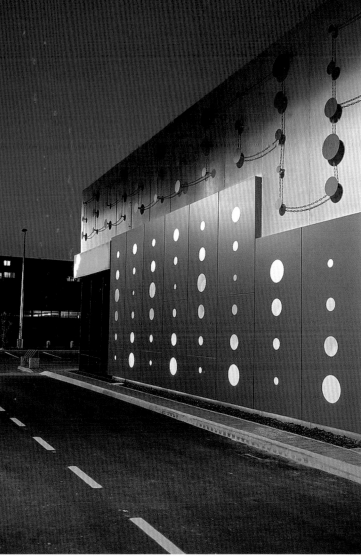

建筑造型和建筑结构

滚动的房子
Roll it Experimental Housing

项目档案

设计：Karen Cilento
项目地点：德国，Karlsruhe 大学

Project Facts

Design：Karen Cilento
Location：University of Karlsruhe, Germany

这是一个很有创意的实验性住宅，是 Karlsruhe 大学学生的作品。这个圆柱形的设计提供了多变的空间。三个不同部分具有各自的功能：床和椅子，运动区域，带有洗涤槽的厨房。
半透明的薄膜包裹着整个表面，上面有赞助商的广告。圆柱的滑动表面是由薄的板条连接构成的，而圆柱的内部是由层压的OSB木板构成的，每一片木板有15毫米厚，圆形的敞口让光线进入圆柱内部。

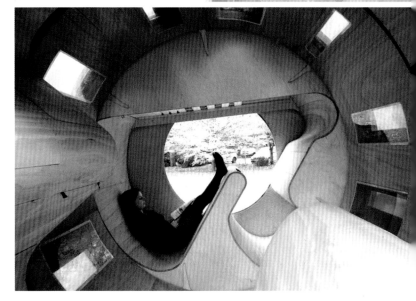

Formative Arts

Roll It, a cool experimental house, resulted from the collaboration among different institutes within the University of Karlsruhe. This cyclindrical design is a modular protype that provides flexible space within a minimum housing unit. Three different sections are dedicated to different functional needs: a bed and table in section, an exercise-ylinder, and a kitchen with a sink.
A translucent membrane envelops the entire form and serves as advertising space for sponsors. Thin wooden slats are attached to the membrane to form the running surface of the roll. The inner cover is a series of laminated OSB panels, each 15 mm thick, which cover the support rings. Circular openings in the side walls allow light into the structure, and a large opening serves as the entrance.

建筑造型和建筑结构

Infomab10 展览亭
Infomab10 Pavilion

项目档案

设计：Kawamura Ganjavian 工作室
项目地点：西班牙，马德里

Project Facts

Design：Studio Kawamura Ganjavian
Location：Madrid, Spain

Kawamura Ganjavian 工作室在预定的时间和预定的预算以内设计了一个很酷的展览亭。这个展览亭是由一个现成的水箱构成的，并且采用了 polyester 加强的纤维玻璃。100 个圆形的小洞让白天的日光进入展览亭。晚上，小孔接收街灯的光线。两个大的开口可以让用户进入展览亭。

这个展览亭原本是一个临时的信息中心，向人们提供最新的公共艺术节目信息；进而转换成为马德里的文化中心。

Infomab10 by Studio Kawamura Ganjavian is a pavilion designed and built in recorded time and within recorded budget. It consists of an off-the-shelf glass-fiber reinforced polyester water tank that was intervened. 100 circular perforations allow speckles of natural light to flood the space during the day, whereas during the night they project the internal light towards the outside like a constellation. Two doors allow circulation through the space.

It was installed temporarily at Paseo de Recoletos in central Madrid as information centre of the public art programme Madrid Abierto, and it was subsequently transported to the Matadero cultural centre.

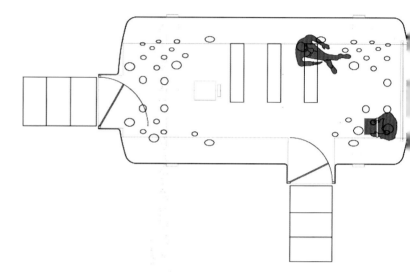

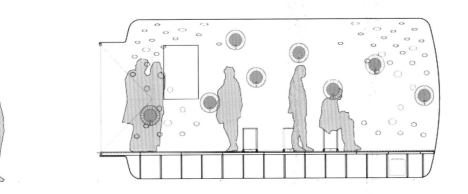
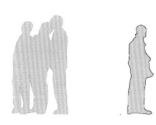

建筑造型和建筑结构

和平大桥
Peace Bridge

项目档案　　　　　　　　　　Project Facts

设计：Santiago Calatrava　　Landscape Architecture：Santiago Calatrava
项目地点：加拿大，卡尔加里　Location：Calgary, Canada

这个项目旨在为人们提供更多的健康的和不影响环境的交通方式选择，提高和完善一个世界级的城市交通枢纽。这座大桥由著名建筑师 Santiago Calatrava 设计指导，整个大桥使用鲜红色的管道形螺旋结构，十分抢眼。

The Peace Bridge meets Council's desire for more sustainable transportation options and supports healthy and environmentally-friendly transportation choices, while enhancing and complementing a world-class downtown core. It embraces the vision set out by Calgarians to create connections between communities and move towards a more sustainable and vibrant city. This bridge is part of an integrated approach to keeping Calgarians on the move and encouraging citizens to use alternate modes of transportation such as walking, cycling, transit and carpooling.

Formative Arts

香奈儿移动艺术展馆
Chanel Mobile Art Pavilion

项目档案

设计：扎哈哈迪德建筑事务所
项目地点：法国，巴黎
面积：700 平方米

Project Facts

Design：Zaha Hadid Architects
Location：Paris, France
Site Area：700 m²

香奈儿700平方米的展馆是Chanel的标志性作品，展馆以极其精湛的细节，共同营造出一个优雅、流畅、连贯的整体。这个多功能多结构的展馆是由一系列连续的拱形结构构成的，中间还有一个庭院。透光吊顶上面的人造灯沿着墙壁直射而下，更加突显了拱形结构，成为另一种新的人造景观。入口处一个大型的采光天窗淡化了室内和室外的差别。除了灯光和色彩效果，空间的节奏感也特别强烈，通过不同部分的接缝，创造出不同的室内视角。中心处的庭院为65平方米，其上方有很多透明的开口。在这里，可以举办各种活动以及为人们提供休息场地。一个128平方米的平台在视觉上与中央庭院联系紧密，继续加强了展馆内部与外部的连贯性。有特殊需要的时候，这两个空间可以合并举行更大型的活动。

The form of the 700m² Chanel Pavilion is a celebration of the iconic work of Chanel, unmistakeable for its smooth layering of exquisite details that together create an elegant, cohesive whole. The resulting functional and versatile architectural structure of the pavilion is a series of continuous arch shaped elements with a courtyard in its central space. Artificial light behind the translucent ceiling washes the walls to emphasize the "arched" structure, and assists in the creation of a new artificial landscape for art installations. A large roof light opening dramatically floods the entrance in daylight to blur the relationship between interior and exterior. In addition to the lighting and color effects, the spatial rhythm created by the seams of each segment gives strong perspective views throughout the interior. The 65m² central courtyard has large transparent openings to the sky above and is designed to host events as well as provide an area for reflection after visiting the exhibition. With a direct visual connection to the courtyard, the 128m² terrace continues the dialogue between the pavilion's exterior and interior. During an event, the two spaces can be linked to become one large event zone.

Formative Arts

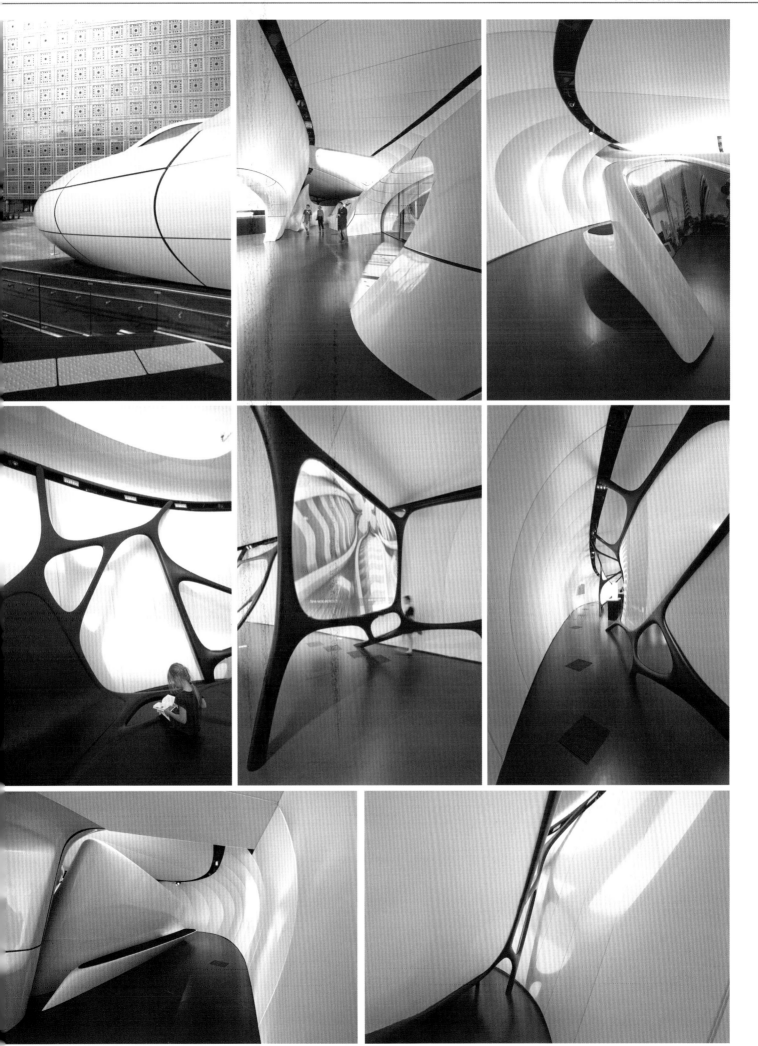

023

雅拉艺术区
Yarra Arts Precinct

项目档案

设计：RWA
项目地点：墨尔本南岸，南岸大马路

Project Facts

Design：RWA
Location：Southbank Blvd, Southbank Melbourne

这个项目包括新的墨尔本剧院和墨尔本演奏厅。剧院和演奏大厅采用钢筋结构，并且具有隔音功能。外立面光滑并采用复杂的管道结构包围着。这个项目挑战性很大，需要和周围的环境相融。

Alfasi Steel Constructions 公司主要负责提供和安装钢结构，Alfasi Design & Drafting 公司主要负责完成 3D 模型和详细的图纸。另外，Alfasi Access Hire 公司提供了长达 30 米的三脚架，并负责安装在一个斜面基地上。RWA 为墨尔本这个新地标设计了外部结构和立面。

The project contains the new home for the Melbourne Theatre Company and Melbourne Recital Hall in the Arts precinct at Southbank. The theatre and auditorium are acoustically isolated steel structures with complex pipe work and a glazed facade. The work requires a high level of project management and coordination with adjoining trades.

Alfasi Steel Constructions is contracted to supply and install the steelworks, while Alfasi Design & Drafting is completing the major task of producing 3D models and detailed steel shop drawing. In addition, Alfasi Access Hire is providing specialist Omme 30m Spider boomlifts for access and installation over a sloping base. RWA provided full design services for all external works for this new Melbourne landmark.

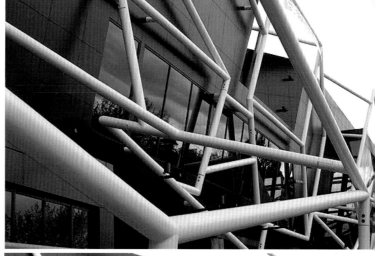
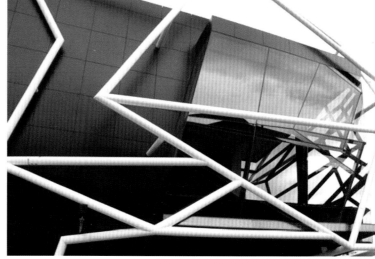

Formative Arts

停车场自动售票机凉亭
House Parking Ticket Machine

项目档案

设计：Jean-Luc Fugier lead architect, FeST Architecture Associates
项目地点：普罗旺斯地区艾克斯 Sisteron 路下车点
面积：30 平方米
完成时间：2010

Project Facts

Dseign：Jean-Luc Fugier lead architect, FeST Architecture Associates
Location：Drop-off parking lot on Sisteron Road, Aix-en-Provence
Site Area：30 m²
Year：2010

设计师创造了一个现代的小型建筑，主要包括一个停车场自动售票机，一个凉亭，一间停车场管理办公室和一间休息室。这个凉亭不仅为这个区域提供了控制便利，而且也吸引了来往人流的注意。考虑到周围的环境以及景观，设计师提出了"小屋"这个设计理念。设计中所选取的材料都来自当地，天然而且对环境没什么影响。从概念到建造，整个过程都是环保的，具有可持续发展性。

Architect Jean-Luc Fugier creates a small modern looking building that currently houses a parking ticket machine, a kiosk window, a parking attendants office and a restroom. The project is located outside Aix-en-Provence, France. The architectural expression of the pavilion is found in the contradiction between controlling and welcoming the public, forming a duality between a plain building that groups the necessary mechanisms for control while realizing that which is necessary to create a welcoming public atmosphere. Driven by the environmental context from which the project is derived as well as the surrounding landscape, the designers looked at the concept of a cabin. Natural and environmentally friendly materials from local industry that are acquired from nearby businesses and carried out by local labour with remarkable skill demonstrate a real approach to sustainable architecture, from concept to construction.

建筑造型和建筑结构

Solearena 休闲椅
Solearena

项目档案

设计：Lutzow 7
项目地点：德国，巴特埃森
完成时间：2010
面积：70 000 平方米

Project Facts

Landscape Achitecture：Lutzow 7
Location：Bad Essen, Germany
Year：2010
Site Area：70,000 m²

该景观是 2010 年德国芭特埃森国际园艺展览会的展览项目之一。它由设计公司 Lutzow 7 进行设计，现在已成为健康疗养度假公园的中心景点。这个圆筒形的结构内部包含了一个温泉，使到访者的神经得到完全放松，充分享受天然矿物盐的益处。

到访者可通过三个管道状的入口的任意一个进入建筑内部。入口仿照木材上的天然纹路而建，建筑外部酷似包着新鲜木材的树皮。建筑内部别有洞天，直接与外部相通。中间是一个温泉，周围摆放着一圈长凳，整个内部形成圆形构造。

Solearena is designed by Luetzow 7 as a part of National Garden Exhibition Bad Essen 2010.
Located within the Solepark, the arena is the focus point. In the arena where brushwood walls surrounded by water, visitors will rest and relax on benches around a fountain basin.

Formative Arts

建筑造型和建筑结构

2011 生态展馆
Eco Pavilion 2011

项目档案

设计：MMX Studio——Jorge Arvizu, Ignacio del Rio
项目地点：墨西哥
完成时间：2011

Project Facts

Design：MMX Studio——Jorge Arvizu, Ignacio del Rio
Location：Mexico
Year：2011

每年，墨西哥都会在 ECO 实验博物馆举行临时展台设计的竞争。今年由 MMX 工作室夺冠。设计没有在主庭院划出一块独立的区域做，反而试图加强与博物馆的联系，创造出一个原有建筑的延伸空间。

在建筑的一角，精心创建一个绳索交织序列，带来丰富的光影和空间变化。这是一个感性的空间，鼓励游客在其间游走探索，可以体验到不同的新领域，新景点，新视角。设计建立了一个全新的体验空间，通过运用绳索交织系统，在院子中创造出不同密度的三维表面，并在局部形成一个较为封闭的空间。空间的开合关系不断变化，并受到绳索投下的阴影影响，在一天中变化无穷，并与游客发生互动，的确精彩无限。

Every year, the ECO Experimental Museum in Mexico City organizes a competition for a temporary pavilion designed to house various events at the main patio of the exemplary building. This year MMX Studio has won the first prize for the 2011 ECO Pavilion. The design does not seek to create a stand-alone piece at the main courtyard; on the contrary, the intervention tries to strengthen the key assets of the original museum, creating an extension of the architectural experiment that the original building pursues.

The intervention encourages the visitors to move around the space and discover new fields, new sights and new perspectives. The new three-dimensional surfaces create screens of varying densities that reconfigure the openness of the original courtyards into a more confined and enclosed space. The new confined space changes constantly as it gets flooded with shadows produced by the rope system. Thus the courtyard becomes an ever changing stage that responds to both the movement of the visitor and the changing patterns of light through the day.

Formative Arts

建筑造型和建筑结构

Aero 展馆
Aero Pavilion

项目档案

设计：Department for Architecture Design and Media Technology
项目地点：丹麦
完成时间：2011

Project Facts

Architects：Department for Architecture Design and Media Technology
Location：Denmark
Year：2011

这个展馆创造了一个循环的气流通道。内部和外部的表面结构与入口垂直，这样，在内部的气流就是向上的，而外面的气流就是向下的。周边的环境可塑性很强，突出了气流的特点，为室内空间增色不少。整个结构是通过数字参数模型建立的，经过简单而快速的生产和组装。这种结构构成部分主要有三种：平行板、垂直板和合板钉。风与灯光融入到结构中去，让简单的结构看似不简单。

The Aero Pavilion creates a circular airflow stream perpendicular to its openings, developing an upward draught at the inner side facing the wind and a downward draught on the opponent side. The invisible environment becomes a constructive element and emphasizes the immediate understanding of the airflow, which again defines the perceptive characteristics of internal space. Environmental conditions of wind combined with the penetration of light through the structure is thereby utilized as means for architectural articulation. The simplicity of the form is created from the controlled complexity of planar plywood plates in digital parametric models for simple and fast production and assembly. Structural integrity is maintained through an interlocking system created of three types of elements-horizontal plates, vertical plates and dowels.

Formative Arts

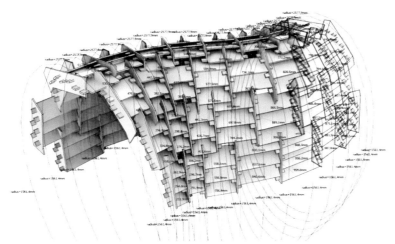

建筑造型和建筑结构

Times Eureka 展亭
Times Eureka Pavilion

项目档案　　　　　　Project Facts

设计：NEX 建筑事务所　　Design：Nex Architecture
项目地点：英国，伦敦　　Location：London, United Kingdom
完成时间：2011　　　　　Year：2011

展亭所在的尤里卡花园在选择植物的时候，注重植物所带来的社会效应，包括在医药、商业和工业方面的用途；特别强调一件重大的事情，那就是我们人类缺乏它们就无法生存，而 Times Eureka 展亭的外观设计便简要地呼应了这样的主题。

NEX 建筑事务所表示，将以密切的角度观察植物细胞结构和成长过程作为设计概念的延伸，是设计过程中不可或缺的方法，最终的结构形式是采取电脑演算法模仿大自然生长的奥妙，并且刻意让到达此处的游客在与日常生活截然不同的空间尺度中，体验生物构造的纹理。这个展亭采用云杉生态林中所采集的木材作为结构，以玻璃作为屋顶材质。展亭的设计集中在对叶片尾管束组织的仿生操作，并将其作为墙体形式，采用木构导管（300dp x140wd）形成基本的形状和支撑结构，次要木构导管则作为支撑板材；专业木材加工厂则在设计团队完成 3D 建模以符合建筑和结构的需求后，着手细部分析结构的数字化生产。

Formative Arts

Plant species chosen for the Eureka Garden reflect their benefits to society including medicinal, commercial and industrial uses underlining the fact we could not survive without them. The pavilion design brief is to reflect the same theme.

NEX Principal Alan Dempsey says that they extended the design concepts of the garden by looking closely at the cellular structure of plants and their processes of growth to inform the designs development. The structural geometry was finalised to use primary timber capillaries (300dp x 140wd) to form the basic shape and supporting structure of the pavilion, inset with secondary timber cassettes that hold the cladding. Following completion of the 3D modelling to meet architectural and structural needs, specialist timber fabricators undertook detailed analysis and digital manufacturing of the structure.

035

南加州建筑学院毕业典礼临时大厅
Graduation Pavilion 2011

项目档案

设计：Oyler Wu Collaborative with SCI-Arc
项目地点：美国，洛杉矶
完成时间：2011

Project Facts

Design：Oyler Wu Collaborative with SCI-Arc
Location：Los Angeles, California, USA
Year：2011

每年的九月，美国洛杉矶的南加州建筑学院都会举行学生毕业典礼，学校的老师和学生会为这一大典准备一个临时展馆。今年，老师 Dwayne Oyler、Jenny Wu 与学生合作，共同设计了这一展馆。展馆共提供 900 个座位，穿越整个停车场并在其北端结束。展馆使用了总长 13 700 多米的绳子、1 800 多米的钢管以及总面积约 300 平方米的遮阳面料，创建出如同树冠般的绳索体系，让遮阳面料漂浮在观众上方向西倾斜，为毕业典礼提供荫凉。在多变的立体构架上采用双层衬垫式网状系统，营造出一个三维的如浪般的遮阳百叶。

展馆基于传统的编织技术，就像织毛衣那样，具有可变性，让编织延展并符合立体造型。与传统的网不同的是，此编织并未在交结处进行固定，这使得网的形状可以在上下范围内能够扭曲。而这种扭曲的特别之处是，当遮阳百叶被其拉伸时会变成跨越这些空间并具有动势的面料，不断地如浪般飘扬飞舞。

项目设计时运用了数字技术及模拟技术进行反复的精心推敲，通过专业的工程公司 Nous，对结构、形状以及网的张力进行持续的评估，以及对网的基本形式进行评估。遮阳百叶阻止了西晒，并为展馆内部提供截然不同的体验。立体双层网深度达到 10'，越往东则变得越开放并且多孔。

Every year in early September, as graduate students at the Southern California Institute of Architecture (SCI-Arc) in Los Angeles put the finishing touches on their thesis projects, a SCI-Arc faculty member and students prepare a temporary pavilion for the annual graduation ceremony. This year, faculty members Dwayne Oyler and Jenny Wu of Oyler Wu Collaborative, along with their students, design a pavilion entitled Netscape for the event that stretches across the northern end of the SCI-Arc parking lot, providing seating for 900. Consisting of 45,000 linear feet of knitted rope, 6,000 linear feet of tube steel, and 3,000 square feet of fabric shade louvers, the pavilion creates a sail-like canopy of rope and fabric that floats above the audience. With its fabric louvers tilted toward the western sky, the canopy is designed to provide shade for the specific date and time.

Netscape utilizes a double layer of netting in varying configurations to create a three-dimensional field of billowing shade louvers. Based on a conventional knitting technique, like that used in the making of a sweater, the pavilion exploits the malleability of this technique as it stretches to conform to the three-dimensional shape of the structure. Unlike a conventional net, the knitting technique is not fixed at its intersections, allowing the shape of the nets (and their grids) to contort both at the upper and the lower surface. With the nets contorting differently, the shade louvers that are stretched between them become a dynamic field of fabric, twisting and bending in order to span across the space in between.

Design of the project involved an elaborate back and forth between digital and analog systems of investigation. With engineering done by Nous Engineering, analysis of the tension in the nets provided constant feedback that informed the shape and three-dimensionality of the structure, as well as some basic form-finding for the nets. As the project progressed, however, large three-dimensional models provided a means of studying the behavior of the grids and their resulting geometries. With the shade louvers designed to block the setting sun in the west, the view from inside the pavilion offers a dramatically different experience. The three-dimensionality of the double-layered netting reaches depths of about 10' and becomes open and porous when facing eastward into the complex three-dimensional field of fabric and rope.

建筑造型和建筑结构

伯纳姆展厅
Burnham Pavilion

项目档案

设计：Zaha Hadid Architects
结构设计：Rockey Structures
灯光设计：Tracey Dear
项目地点：美国，芝加哥

Project Facts

Design：Zaha Hadid Architects
Structural Engineers：Rockey Structures
Lighting & Electrical：Tracey Dear
Location：Chicage, USA

展厅坐落于密歇根湖千禧公园中央的一处独特的地方，一侧紧临密歇根大道。项目的本意是想吸引更多的人前来散步、游览和观光。基于对城市背景、项目类型及规模的考虑，展厅设计既体现了现有规则的几何构造，又营造出富于动感和多维的空间效果。它很好地适应了城市环境和千禧公园的客流，引入了面向公园和城市环境的诸多景观。

展馆的外部轮廓是铝制金属，并装有彩灯。在夜晚，它就像一个灯塔，闪耀着各种缤纷色彩。作为伯纳姆规划百年庆典的一部分，这个展厅用它的外观展示了芝加哥城的轮廓。

Zaha Hadid Architects's pavilion merges new formal concepts with the memory of Burnham's bold, historic urban planning. Superimpositions of spatial structures with hidden traces of Burnham's Plan are overlaid and inscribed within the structure to create a dynamic form. The Burnham Plan Centennial is all about celebrating the bold plans and big dreams of Daniel Burnham's visionary Plan of Chicago. It's about reinvention and improvement on an urban scale and about welcoming the future with innovative ideas and technologies. The designers' design continues Chicago's renowned tradition of cutting edge architecture and engineering, at the scale of a temporary pavilion, whilst referencing the organizational systems of Burnham's Plan. The structure is aligned with a diagonal in Burnhams early 20th century plan of Chicago. The designers then overlay fabric using contemporary 21st century techniques to generate the fluid, organic form, while the structure is always articulated through the tensioned fabric as a reminder of Burnham's original ideas.

建筑造型和建筑结构

Hexigloo 展厅
Hexigloo Pavilion

项目档案

设计：Tudor Cosmatu, Irina Bogdan, Andrei Raducanu
项目地点：罗马尼亚，布加勒斯特

Project Facts

Design：Tudor Cosmatu, Irina Bogdan, Andrei Raducanu
Location：Bucharest, Romania

这是一个全部由参数化设计而成的展厅，设计基于蜂窝状的单元小结构，呈现出小圆屋的形状。在预先制作好的行列数各为14的表面上画上六边形，并将在Z轴线处的六边形挤压出来，最后在整体结构上增加硬度。设计的焦点在于室内空间，圆锥形的漏斗将光线带入室内，当进入结构内部时，光滑的室外和精致的室内展现出强烈的对比和反差，给人极大的惊喜。

Hexigloo is a pavilion based on the cellular structure of a honeycomb applied to a igloo surface typology. From concept to the final product, the process went through following steps: mapping a hexagonal grid on a pre-modeled surface (14 rows + 14 columns, resulting in 196 elements), extruding the mapped hexagons on the Z axis in order to create a binding surface between the components and finally adding rigidily to the overall structure. The main focus was on the interior space, which was articulated by cone like funnels that filter light into the interior space. When entering the structure, the contrast between the smooth exterior and the intricate interior reveals a moment of surprise.

Formative Arts

建筑造型和建筑结构

Tverrfjellhytta 野生驯鹿观景亭
Tverrfjellhytta

项目档案

设计：Snohetta Oslo AS
项目地点：挪威
面积：90 平方米
完成时间：2011

Project Facts

Design：Snohetta Oslo AS
Location：Hjerkinn, Dovre Municipality, Norway
Site Area： 90 m^2
Year：2011

这座观景亭占地 90 平方米，位于挪威国家公园内，这里是野生驯鹿生活的地方。建筑位于海拔 1200 米的高原上，不仅为游客提供了绝佳的视野，而且还是野生驯鹿基金会教育项目的观摩亭。

独特的自然和文化景观是这座建筑的设计源泉，建筑设计是基于一个坚固的外壳和有机的内核。建筑师把重点放在能抵御恶劣天气的材料选择上，长方形的框架采用了类似于铁的粗钢材质，简洁的造型和天然材料的选择符合当地的建筑传统。南立面创造了曲线形的座椅，使人们在上面休息时沐浴在温暖的阳光之中。

建筑师在建造过程中结合挪威的造船技术和先进的制作工艺，每 0.9 平方米的木质横梁都被打磨光滑并且用木楔进行加固，创造出波浪般的立面效果。外墙使用松焦油进行处理，内部木材也经过油化加工。坚硬的长方形外框与生锈的钢材相结合，自然而然地融入到周边环境中，同时对外部木板的特殊处理，也突出了高山徒步者的存在。

The Norwegian Wild Reindeer Centre Pavilion is located at Hjerkinn on the outskirts of Dovrefjell National Park, overlooking the Snohetta mountain massif. The 90m^2 building is open to the public and serves as an observation pavilion for the Wild Reindeer Foundation educational programmes. A 1,5km nature path brings visitors to this spectacular site, 1,200 meters above sea level.

This unique natural, cultural and mythical landscape has formed the basis of the architectural idea. The building design is based on a rigid outer shell and an organic inner core. The south facing exterior wall and the interior create a protected and warm gathering place, while still preserving the visitors view of the spectacular panorama. Considerable emphasis is put on the quality and durability of the materials to withstand the harsh climate. The rectangular frame is made in raw steel resembling the iron found in the local bedrock. The simple form and use of natural materials reference local building traditions. However, advanced technologies have been utilized both in the design and the fabrication process. Using digital 3D-models to drive the milling machines, Norwegian shipbuilders in Hardangerflord created the organic shape from 10-square-feet square pine timber beams. The wood was then assembled in a traditional way using only wood pegs as fasteners. The exterior wall has been treated with pine tar while the interior wood has been oiled. The pavilion is a robust yet nuanced building that gives visitors an opportunity to reflect and contemplate this vast and rich landscape.

Formative Arts

建筑造型和建筑结构

壁炉式儿童游乐场
Fireplace for Children

项目档案

设计：Haugen/Zohar 建筑事务所
项目地点：挪威

Project Facts

Design：Haugen/Zohar Arkitekter
Location：Trondheim, Norway

因为预算非常有限，所以设计师只能使用周围建筑工地中的剩余材料来建造这个户外壁炉，因此建筑的设计以短木件为基础。这个建筑的设计灵感来源于挪威的一个草坪小屋和旧式的原木建筑，它的尺寸为 5.2 米 X4.5 米。木制结构由 80 个圈组成，每个圈由 28 片松木构成，松木与松木之间形成了自然的缝隙，自然的光线透过这些缝隙进入到空间内部，整个木制结构看起来像是一个烟囱，户外壁炉还有一个双层的曲线形滑动门。

Given a very limited budget, reusing leftover materials (from a nearby construction site) was a starting point that led the design to be based on short wooden pieces. Inspired by the Norwegian turf huts and old log construction, a wooden construction was built and mounted on a lighted and brushed concrete base. The structure is made of 80-layered circles. The circles have varied radiuses and relative center point in relation to each other. Every circle is made out of 28 pieces of naturally impregnated core of pine that are placed with varied spaces to assure chimney effect and natural light. Oak separators differentiate vertically between the pine pieces to assure airflow allowing easy diying of the pine pieces. A double curved sliding door is designed for locking the structure.

温尼伯湖冰上庇护所
Winnipeg Skating Shelters

项目档案	Project Facts
设计：Patkau 建筑事务所	Design: Patkau Architects
项目地点：加拿大，温尼伯湖	Location: Winnipeg, Manitoba, Canada
完成时间：2011	Year：2011

Patkau 建筑事务所为滑冰者设计了一组位于加拿大温尼伯的庇护所，目的在于协助滑冰者在休息时规避冬季寒冷刺骨的寒风。庇护所由一所有机的圆锥外形小屋组成，小屋都是由两层易弯的胶合板组成，胶合板特有的弯曲特性产生了足够的强度，与木质框架共同支撑着小屋。内部木质的地板与座椅带给使用者温暖而舒适的感觉，而且为小屋增添了很多空间和结构上的特色。

每个木屋都包括一个三角形的地基和边缘经过处理了的木脊，这个木脊可以减轻积雪对木屋的压力。小屋和小屋之间的位置关系看似是没有任何规律的，其实不然，两个木屋之间经过 120 度的旋转，第三个小屋则有 90 度的旋转，这样就构成了小屋和小屋之间的内部空间。

The program has developed to sponsor the design and construction of temporary shelters located along the skating trails. Each shelter is formed of thin, flexible plywood which is given both structure and spatial character through bending and deformation. Skins, made of 2 layers of 3/16th-inch-thick flexible plywood, are cut in patterns and attached to a timber armature which consists of a triangular base, and wedge shaped spine and ridge members. The ridge is a line to negate the gravity loads of snow.

Grouping the shelters into a cluster begins with the relationship of two, and their juxtaposition to qualify the size and accessibility of their entrance openings. This apparently casual pairing is actually achieved by a precise 120-degree rotation. Three pairs (one with mirror reflection) are then placed in relation to one another through a secondary rotation of 90 degrees to form the cluster and define an intermediate "interior" space within the larger grouping.

Formative Arts

047

建筑造型和建筑结构

"明暗"投影咖啡店
Shading Surface

项目档案

设计：Ateliermob
项目地点：葡萄牙，里斯本

Project Facts

Design：Ateliermob
Location：Sacavem, Portugal

该项目位于葡萄牙里斯本郊区Sacavem镇的一条道路上，属于城市限制性规划。项目涉及到三个问题，现在只解决了两个。拟建楼宇有幸成为公共场所，在建筑物之间起到决定性作用，而在这之前这些建筑物都是准备遗弃或是没有公共的资格的。通过实地考察和深刻分析城市规划的动态性，该项目决定建造三个同本质但是不同功能的部分，并且，这三个部分需要简化周围复杂的环境，减少对周围环境的影响。最终，三个立面不同，视野不同的安静部分就此诞生。斜屋顶结构丰富了室内设计的语言，增强了天花板和地面的张力。

This intervention is part of an urban requalification plan for one avenue at Sacavem on the outskirts of Lisbon, promoted by Loures City Hall. The project held three interventions, but only two were completed. The proposed buildings had the common goal of playing the decisive role at the qualification of public spaces between buildings, formerly abandoned and disqualified. After the initial site visit and interpretation of the urban plan's main dynamics, it was felt the need to design three built elements sharing a common nature, although holding different uses and problematics. Therefore, the designers sought to design three quiet buildings under the exciting motto——"nothing new under the sun". The solution to the problem seemed to be in the so called fifth elevation (roof plan) that emerges as the decisive factor in the solution of the elevations differences and the visual extension of the street facing north. After some volumetric studies, and reflecting the will of the acquaintance of these three interventions as a building compound, the designers have chosen to go for a common solution of three roof slopes structures carrying a huge dramatic potential in their interior spaces and a continous tension between the ceiling and the floor.

Formative Arts

049

建筑造型和建筑结构

Calder Woodburn 休息区
Calder Woodburn Rest Area

项目档案　　Project Facts

设计：BKK 建筑事务所　　Architects：BKK Architects
项目地点：澳大利亚　　Location：Australia
面积：549 平方米　　Site Area：549 m²

在澳大利亚，这个休息区域有着长久和丰富的历史。大面积的漂浮状屋顶给每一位路人都留下了深刻的印象，不仅可以挡风遮雨，而且本身亦是一个建筑标志。这里不仅可以让路人得到暂时的休息，而且还能够让路人在旅途中停下来思考。

博物馆在城市和城镇中到处可见，它是文化与历史的承载者。这个项目不仅仅是一个休息区，也不仅仅是一个洗手间，而是一个能鼓励人们阅读和思考的博物馆。这个休息区可以通向 Calder Woodburn 纪念大道和大型的谢珀顿休息区域。在休息区的旁边，有一系列的交通道路和植物，还有一个信息板。路人通常都会驻足来阅读信息板上的信息。

洗手间的建造采用标准的道路建造技术——在加固的混凝土路面上安置预先浇灌的混凝土板块。这样不仅可以节省施工时间，而且可以减少人力和建筑成本。屋顶所采用的材料是钢筋，室内的隔板是用彩色的砖块砌成的，外面则用混凝土包裹，成贝壳状。总之，所有的材料在选择的时候都注重较强的稳固性和较低的维护成本。休息区的座椅在设计的时候注重流线性，也注重视野的开阔性。

Formative Arts

建筑造型和建筑结构

Formative Arts

There is a long and rich history of the service station within Australia. The familiar image of the large floating roof providing shelter to the amenities of the service provider is a strong memory for all who have undertaken a road trip. It performs a vital role in the rest and re-supply needs of the traveler. These places mark a point in a journey, a place for pausing, a place for reflection. The roof serves as both shelter and icon, signaling to the passers-by while serving the practical function of sheltering.

Monuments are an integral and familiar part of cities and towns. They provide a valuable insight into the culture of a place and locate it within a broader history. Our building is placed on a plinth that raises it out of the flood plain and gives the structure a civic monumentality that encourages readings and understandings loeyond the notion of the "toilet block". This building is to be read as a gateway to the Calder Woodburn Memorial Avenue (CWMA) and also to the larger Shepparton Area. The CWMA naming plaque has been relocated amongst a landscaped system of paths and planting adjacent to the rest station. Visitors will pause to read the plaque against the backdrop of planting that repeats the Avenue of Honour as they make their way to the picnic tables.

The construction of the toilet blocks themselves utilises standard road building techniques of pre-cast concrete units located on a reinforced concrete slab. This process minimises on-site time, labor and construction costs. The steel-framed roof is then built over these units to provide shelter. The interior partitions are constructed from brightly coloured glazed bricks providing a jewel-like visual contrast to the internal concrete shell. Materials have been chosen for their robustness and low maintenance. The siting of the structures is designed to emphasise the linearity and contained views of the CWMA.

建筑造型和建筑结构

迈阿密帐篷
Design Miami/ Tent / Moorhead

项目档案

设计：Moorhead & Moorhead
项目地点：美国，迈阿密

Project Facts

Design：Moorhead & Moorhead
Location：Miami, USA

设计以超越传统的全封闭式帐篷概念，突出帐篷骨架，在仅有的结构范围内打造出新的空间和概念，最后设计出以最常用的帐篷建造材料打造出最出乎意料最新奇的帐篷。

为了建造这个临时的结构，建筑师们与制造商紧密合作，最终的设计很好地抓住了帐篷的本质，不会隐藏任何帐篷中的结构。这样不仅满足了帐篷的商业功能，还以全新的空间和深度为人民提供了新的体验。空间裸露的部分采用白色油漆粉刷，给人一种温暖的感觉。顶棚的立面是由乙烯基构成的，并且只是部分覆盖住，通过各种折叠和截断，创造了一种马赛克样式，这种样式会随着灯光和周围的环境而呈现出动态的美。

Formative Arts

Commissioned by Design Miami to transcend the appearance of a traditional clear span tent while working within the limitations of the structure, Moorhead & Moorhead's final design manipulates the most common tent materials in unexpected and innovative ways.

To develop the temporary structure, Moorhead & Moorhead worked closely with manufacturer EventStar, the leading expert in the field. Together, their final design embraces the true nature of the tent by never camouflaging the design elements typical of tent structures. Simultaneously, the design elevates the idea of commercial tent design in a way that will delight visitors while subverting their sense of space and depth. The final design includes a whitewashing of the space's exposed elements, creating a welcome sense of stillness and respite within its buzzing South Beach setting.

The design of the temporary structure will feature a partially covered courtyard in which guests will be allowed to linger and relax. Moorhead & Moorhead have created this space by folding and cutting the tent's vinyl facade in a striking mosaic pattern that will create ever-changing displays of light and shadow.

顶棚和亭子
A Canopy and a Pavilion

项目档案

设计：Matthieu Gelin & David Lafon
项目地点：法国，巴黎
灯光设计：Light Cible
面积：250 平方米

Project Facts

Design：Matthieu Gel in & David Lafon
Location：Paris, France
Lighting：Light Cible
Site Area：250 m²

这个白色的亭子从地铁出口延展开来，成折叠状。不仅具有动态美，而且特色鲜明，能够立即吸引游客的注意。这个亭子除了具有特别的艺术感，功能还很强大，能够容纳两辆巴士。亭子结构和材料的选择保证了亭子的稳固性。从地铁入口延伸出来的折叠面不仅是一个顶棚，而且还可以作为一堵墙。顶棚下面的柱子散乱地排列着，这个公共空间可以用作巴士的停靠点，地铁入口和单车的停靠点等。这个亭子在材料的选择上注重与周围环境的和谐统一，主要采用白色的钢筋，轻巧而稳固。亭子整体看起来就像是大海中的帆布，让周围的建筑也有了一种动态之美。

The pavilion is a white folded shape that emerges from the ground at the entrance of the subway station Porte des Lilas. This monolithic shelter provides a dynamic shape and strong signal for travelers. Exceptional by its great dimension, its sharpness and folding are both artistic and functional, appropriately providing enough space for two buses.
The shelter is a pure construction. Its expression is linked to the elements of structure that compose it and which are necessary to its stability. The canopy holds its strength and continuity through materials. The folding that comes out from the ground forms both a wall and coverage. Pilars stroll freely like pedestrians in a group. The pavilion provides helps in establishing the global functioning of this public place: bus terminus, subway entrance, shelters for free service cycles, meeting area for the cinema and the circus.

Formative Arts

Its grip on the ground, very lightweight, enables a continuity and transitions through its total use of materials including the paving stones. The shelter appears more as a sail delicately put in the limits of the square rather than an immovable building. The whole shelter is constructed with one material——steel painted white.

建筑造型和建筑结构

聚会的地方
The Meeting Place

项目档案

设计：ASPECT 工作室，Herbert & Mason, Serlot Studio
完成时间：2010
所获奖项：2010 年 AIAL 新南威尔士设计奖
项目地点：澳大利亚，悉尼，小猎人街

Project Facts

Design：ASPECT Studio, Herbert & Mason, Derlot Studio
Year：2010
Awards：2010 AILA NSW Awards——Design
Location：Little Hunter Street, Sydney, NSW, Australia

这个项目部分是建筑外部设施安装，部分是雕塑和社会试验。这个设计极具创意和乐趣，在鼓励人与人之间的参与和互动的同时，丰富了人们穿过城市小街道的体验。

设计概念主要是在一个现有的过道里构建两个 4 米高的有弹性的幕墙。幕墙是不透明的，两个幕墙之间留有缝隙，两边的街景不受干扰。人们穿过这里时，如果碰巧对面有人过来，那么就需要和对方交流和妥协。在这个过程中，社会中人们之间的观察，交流和协商以一种有趣的方式得到体现和促进。

"The Meeting Place" is part architectural installation, part sculpture and social experiment. It is a playful design which encourages participation and interaction whilst heightening the experience of moving through the urban space of little Hunter Street.

The concept is to create a space within the existing lane way by creating two 4m high curtain walls of elastic fabric. The material has an opacity to it which allows views through and when lit at night becomes a canvas for revealing movement of people through the space. People have to negotiate their way through the laneway by communication and contact with other people like themselves who are moving in the opposite direction. This social aspect of watching, communicating and negotiating with people will increase positive human contact with a sense of play.

建筑造型和建筑结构

Formative Arts

建筑造型和建筑结构

USCE 购物中心
USCE Shopping Centre

项目档案

设计：Chapman Taylor Architetti
项目地点：塞尔维亚，贝尔格莱德

Project Facts

Design：Chapman Taylor Architetti
Location：Belgrade, Serbia

USCE 是塞尔维亚最具有活力和现代化的零售和休闲购物中心，位于贝尔格莱德中部，这个购物中心包括 120 多家商店，另外还有一个电影院、一个保龄球馆、一个俱乐部、一个家庭娱乐区域。总面积为 130 000 平方米，包括三层商业楼层和两层停车场。

这个项目的地理位置很重要。项目的后面有一个大型的城市公园，位于两条河流的交汇处，连接着贝尔格莱德的新区和旧区。城市公园正对一个大型的三角形购物大楼，购物中心立面主要采用锌、钢铁和玻璃，为这个城市创造了一个新的地标，并且与周围区域的景观地形高度协调统一。

建筑的风格在晚上更加引人注目。灯光从不同角度照射在建筑上，形成不同的阴影和效果。

USCE is Serbia's most innovative and modem retail and leisure destination. Situated in the heart of Belgrade, it has more than 120 shops, as well as a cinema, bowling alley, casino and a family entertainment area. The 130,000m^2 centre covers three commercial floors and incoporates two levels of car parking.

The building is located in an area of great importance to the city, backing onto a large urban park where the Sava and Danube rivers meet, and acts as a strategic connection between the new and the old Belgrade. Fronted by a large triangular plaza, the main facade of zinc, steel, and glass creates a modem landmark for the city with striking and sculptured elements which are echoed in the landscaping of the surrounding area, characterised by strong geometric patterns.

The building's innovative architectural style is emphasised in the evenings through the lighting that illuminates the facade from different angles with different shades and degrees of intensity.

Formative Arts

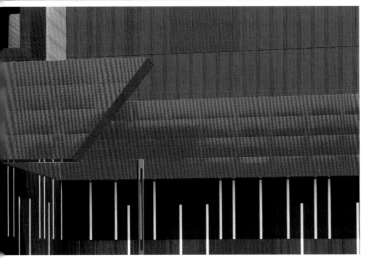

建筑造型和建筑结构

Grafenegg 展馆
Grafenegg Pavilion

项目档案

设计：Land in Sicht, Thomas Proksch
项目地点：奥地利
灯光设计：Christian Ploderer, Ploderer & Partner

Project Facts

Landscape Architecture：Land in Sicht, Thomas Proksch
Location：Grafenegg, Austria
Lighting Design：Christian Ploderer, Ploderer & Partner

这个区域总面积有31万平方米，中间有城堡和壕沟。Grafenegg 的城堡有250年的历史，每一个时期都留下了岁月的痕迹，这也是这个公园吸引人的原因之一。公园内一年四季苍翠，在东面和西面有两个主要的入口。敞开式的展馆在夏天成为举行重要盛会的地方，在平时也为游客提供休息的地方。

The castle grounds of Grafenegg are almost 250 years old. Every period has left its traces, and this stylistic variety is one of the reasons for the park's appeal. The area, with the castle and moat at its centre, has a size of about 310,000m². The park and its dendrological collection are open year round and are accessible from two main entrances on the West and East side. The open-air pavilion to be erected in the park is used as a stage during festival season in summer, and as an attraction for excursionists and flaneurs, similar to the gazebos in historical landscape gardens, which were designed as a destination or a stop-over on extended walks.

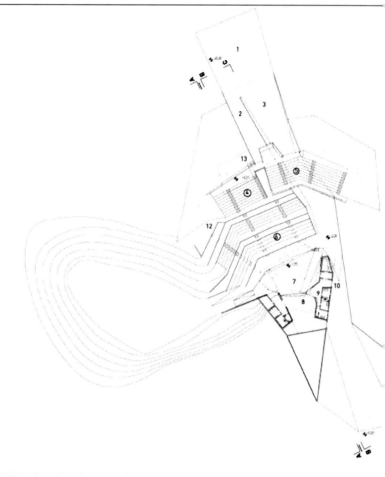

Formative Arts

065

建筑造型和建筑结构

洛尔卡广场
Lorca's Square

项目档案	Project Facts
设计：Jesus Torres Garcia	Design：Jesus Torres Garcia
项目地点：西班牙	Location：Salobrena, Spain
面积：1 800 平方米	Site Area：1,800 m²

洛尔卡广场整体均由混凝土建造。混凝土必须具备最典型的素质，才可能在整个城市中运用，比如一种像一层薄膜的过滤混凝土。刚开始项目组打算采用一种蓝色添加剂，但由于成本昂贵而放弃。广场上由玻璃制成的雕塑最后成型的是一张上面刻有些许文字的弯曲的薄板，白板黑字，成为萨各夫雷纳的孩子的"迪斯科舞厅"、大学出口的一个隐蔽处、城市中心的一处安全的所在，就好像田地中的房屋，是第一个孩子们能触摸得到的地方。

Lorca's Square was done with concrete of some exemplary qualities to extend it in all the city. We discovered a blue additive that was very much successful in the beginning, but the additive was very expensive, ultimately we decided to give up. The sculpture was of glass, and finally it ended to be a humble curve sheet with some inscriptions, in black and white, for the children of the city of Salobrena, a "disco for kids" a hiding place to the exit of the college, a place in which to protect itself in the middle of the city, as the huts in the field, the first space that the children can feel.

Formative Arts

067

建筑造型和建筑结构

罗马尼亚 ZA11 亭子
ZA11 PAVILION in Romania

项目档案

设计：Dimitrie Stefanescu, Patrick Bedarf, Bogdan Hambasan
项目地点：克鲁日，罗马尼亚
完成时间：2011

Project Facts

Design：Dimitrie Stefanescu, Patrick Bedarf, Bogdan Hambasan
Location：Cluj, Romania
Year：2011

这是一个很有雄心的项目，由学生们集体设计，并按照1:1的比例专为罗马尼亚克鲁日ZA11话说建筑活动建造。这个亭子极好地融入周围的历史环境，同时呈现出强烈的代表性。充满象征性的设计风格满足了亭子的主要设计目标：吸引周围的路人。这项工程尝试着将新的本体论变得清晰。本体论在计算式建筑风格的规定下，展现了亭子设计的不同进程。同时，亭子为在建筑周围举办的不同社会活动提供了一个纳凉的地方，也可以用来庆祝相应的建筑节日。

在设计的过程中，设计师面临一个很严峻的考验，那就是在使用赞助商提供的材料和工具设计出可行的施工设计的同时，还要按照预期的那样，尽量降低成本。因此设计师对相对有限的方法进行了创造性的探索，更为重要的是，要在现有的材料和建筑技巧下，找到可行的方法。最终设计包含746个单独的设计，将它们组装起来可以形成一个自由形式的圆环，圆环可以进行进一步的划分，形成六边形结构。这种特殊的几何形状，使亭子可以成为众多活动的举办场所，同时以其特别的外形吸引众多的游客。

如果没有那么多志愿学生的帮忙，这个项目不可能那么顺利和成功。各种各样厚度的材料，后续那么多的不确定性，较少的接头要求，以及风和雨这些问题都需要及时解决。作为一个很好的学生教育实践，学生们在这个过程中充分理解和处理现实的一些限制因素，创造出一个无价的亭子。

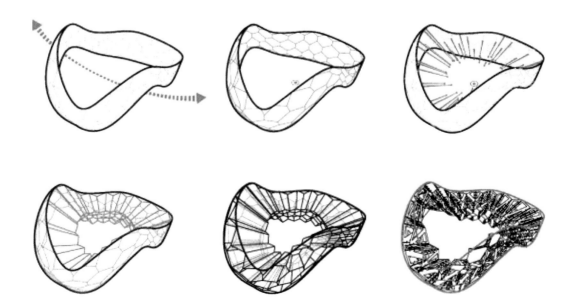

Formative Arts

建筑造型和建筑结构

Formative Arts

The project started out as an ambitious student-powered endeavor to design and fabricate at a 1:1 scale the flagship pavilion for the ZAU Speaking Architecture event in Cluj, Romania. While at the same time integrating into its historically-charged context, the design boasts a strong representational power which is much needed in order to fulfill its main goal of attracting passers-by to the event. The object tries to make legible the new ontology which is slowly defined by computational architecture and is a showcase for the processes empowered by it. At the same time, the pavilion offers a sheltered space for the unfolding of different social events pertaining to the corresponding architecture festival.

The design was elaborated during a parametric design workshop specifically geared towards its production. The designers were faced with the harsh requirements of creating an actually working design with the material and tools available from sponsors while at the same time fitting inside a budget dwarfed by its expectations. Therefore the designers constrained the creative exploration agenda to a relatively limited approach which, most importantly, was scalable in terms of materials and fabrication techniques. The final design consists of 746 unique pieces, which, once assembled, create a free-form ring which is subdivided into deep hexagons. This particular geometrical configuration allows for the sheltering of the different planned events while at the same time inciting curiosity through its unusual, spectacular form. The realization of the design is made possible by advanced use of parametric design techniques, with the help of which the whole process was controlled from exact geometry generation to piece labeling, assembly logic and actual fabrication (CNC milling).

The actual assembly process wouldn't have been possible without the team of students which volunteered to help. As an educational exercise, it completed the design phase and proved to be invaluable in terms of actually understanding and working with the constraints encountered in real-life. Varying material thickness and subsequent extra flexibility and less joint stiffness, rain and wind posed many challenges which had to be resolved on-site as quickly as possible so as to meet the assembly deadline.

绿色小酒馆室内设计
Green Bistro Interior Design

项目档案

设计：Siddik Erdogan & Jorn Frohlich
项目地点：德国
面积：100平方米
完成时间：2011

Project Facts

Design：Siddik Erdogan & Jorn Frohlich
Location：Osnabruck, Germany
Site Area：100 m²
Year：2011

这间德国小酒馆的室内设计是由 Siddik Erdogan & Jorn Frohlich 设计的，设计的理念围绕着健康的生活方式以及绿色饮食而展开。因此，设计团队在装修设计上不选用任何带有锋利边缘的形状（如长方形等）。材料多选用白色漆面的胶合板。所有的功能部件，如食物展示架、托盘、桌子都被考虑到整体设计中。

为了增加接触的感官体验，白色的椅子和样式化格子被安装其中。设计的视觉核心是创建了一个可以覆盖100平方米的独特天花板，一个波浪形的独特装置被安装在吊顶上。

Formative Arts

The basic concept evolves around the lifestyle of health and sustainability promoting healthy green, food items to the customers. The choice of material reflects that idea. As for design inspirations, the team chose organic shapes derived from nature itself avoiding any kind of rectangular forms or sharp edges. The materials consist of naturally enhanced plywood combined with white lacquered surfaces. All functional parts such as food displays, tray shelves, benches and tables have been merged into overall organically shaped objects that have been designed, planned and manufactured individually.

In order to add a sophisticated touch to the organic design idea, white pantone chairs and white tulip tables have been installed. To complete the team's conceptual vision, a ceiling centerpiece covering all 100 m² of the bistro area has been created.

电视秀凉亭
Gazebo for TV Show

项目档案

设计：za bor Architects/Arsenity Borisenko
项目地点：俄罗斯
完成时间：2011

Project Facts

Design：za bor Architects / Arseniy Borisenko
Location：Russia
Year：2011

这个项目是为参加热门电视节目《乡村会谈》而做的。参与这个节目的观众的别墅或者住房将由电视台邀请的建筑师和设计师来改建。参与者不用付费，同时也不能主导任何结果，所以，面对建筑师的纯粹想法，参与者总是惊喜连连。

这个项目位于典型的郊区中，业主喜欢烧烤以及园林植物。于是，建筑师用白色的落叶松支撑了14个面，并将它们拼起来，形成一个像海浪似的、螺旋状的自支撑结构，里面有用餐区和烧烤区。这里被四季美丽的色彩所包围，夏季是葱郁的绿，秋天是绚丽的黄和红。整体透明开放，激发了人与自然的关联。

项目的主要建筑师 Arseniy Borisenko 和 Peter Zaytsev 表示："我们想做出一个复杂的、动态的结构，不光具备功能性，还可与现有场地保持协调。虽然这个结构有14个面组成，形体复杂，但是我们采用了中性的色彩以及天然的落叶松材料。这些既强有力地展现了项目本身，又融合在场地之中，放眼皆是美。这个构筑物还是遮风避雨的好地方，我们希望业主除了夏天，冬季也可以使用这里。"

The project has been developed specially for popular TV show "The village talks". The idea of the show is that for those owners of country houses and cottages who agreed to participate in the experiment, the invited designers or architects do replanning of a part of their village. The important moment is that the house owners pay nothing for reconstruction, but at the same time they can't influence the result, so it always comes unexpected for them. The architects in their turn try to offer the most original solutions.

The object here is in fairly typical suburban area, with garden trees belonging to the captain of the yacht, who enjoys cooking on the grill with his family and a number of friends. Architects have suggested making a small-size self-supporting structure consisting of fourteen planes made of larch white-tinted wood. The gazebo has the helical structure resembling a sea wave, with an area for feasts and a barbecue area with a chargrill made of brick and steel. Architects, concept for the construction is transparency and openness which inspires a contact between man and nature, especially because of surroundings: a green lawn and wonderful fruit trees giving an abundant harvest each fall. Neutral tints of the gazebo are drowning in intense colours of the garden from the lush green in summer to yellow and red in autumn, and bringing together a rather complex and aggressive form with pastoral Moscow suburbs, allowing it to exist peacefully within the site context.

Arseniy Borisenko and Peter Zaytsev are making comments on the project: "We wanted to develop a complex dynamic structure that would not only perform its functions-gazebo and chargrill area, but would preserve the existing context of the site.

Formative Arts

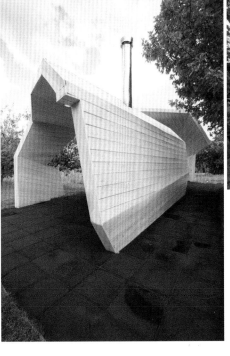

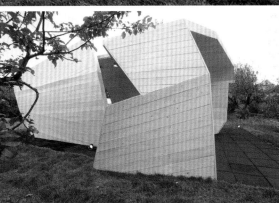

Although our project is a complex structure consisting of 14 flat segments, we used neutral colours and natural larch wood. This helps, on the one hand, to present an object effectively and emphasize its structural features, on the other hand, to leave it in the existing suburban context, to fuse in the greenery of the garden, to please the eye, not to offend it. The gazebo planes are an excellent protection from the wind and rainfall, so we hope its new owners will be able to use it not only in summer but in winter as well".

日落教堂
Sunset Chapel

项目档案	Project Facts
设计：BNKR Arquitectura	Architects：BNKR Arquitectura
项目地点：墨西哥	Location：Acapulco, Guerrero, Mexico
面积：120 平方米	Site Area：120 m²
完成时间：2011	Year：2011

这个项目主要有两个设计目的：首先是希望这里可以作为新婚夫妇新生活开始的地方；其实是希望可以在这里寄托对亲人的思念。完全相反的两个目的似乎成为了这个设计的主要驱动力。前者赞美生命，后者哀悼死亡。对立的要素开始碰撞：玻璃与混凝土；透明度与坚实度；虚无与沉重；古典比例与混沌外观；脆弱与坚不可摧；短暂与持久。设计要求直白简单：首先要充分利用教堂外壮丽的景色；其次，太阳与祭坛的关系要求准确；然后最后一个要求，建筑要融入场地。业主期望这是一个能与天体运动周期发生关系，乌托邦式的完美建筑。

场地中有两个主要的因素阻挡视线：大面积的丰富植被以及庞大的巨石。为了避开这些不利因素（搬开大石头是不可能的，从伦理、精神、环境、经济各方面原因综合考虑），教堂至少需要抬高 5 米。这里是一个异域风情，美景如画的处女地，建筑师努力将接地面的面积缩小，减少对场地的影响，并将建筑处理得如同场地中的石头。阿卡普尔科山由巨大的花岗岩形成，建筑里力求将教堂做成"山顶上的巨石教堂。"

The designers' first religious commission was a wedding chapel conceived to celebrate the first day of a couple's new life. Their second religious commission had a diametrically opposite purpose: to mourn the passing of loved ones. This premise was the main driving force behind the design. The two had to be complete opposites, and they were natural antagonists. While the former praised life, the latter grieved death. Through this game of contrasts all the decisions were made: Glass vs. Concrete, Transparency vs. Solidity, Ethereal vs. Heavy, Classical Proportions vs. Apparent Chaos, Vulnerable vs. Indestructible, Ephemeral vs. Lasting. The client brief was pretty simple, almost naive: First, the chapel had to take full advantage of the spectacular views. Second, the sun had to set exactly behind the altar cross (of course, this is only possible twice a year at the equinoxes). And last but not least, a section with the first phase of crypts had to be included outside and around the chapel. Metaphorically speaking, the mausoleum would be in perfect utopian synchrony with a celestial cycle of continuous renovation.

Two elements obstructed the principal views: large trees and abundant vegetation, and a behemoth of a boulder blocking the main sight of the sunset. In order to clear these obstructions (blowing up the gigantic rock was absolutely out of the question for ethical, spiritual, environmental and economical reasons), the level of the chapel had to be raised at least five meters. Since only exotic and picturesque vegetation surrounds this virgin oasis, the designers strived to make the least possible impact on the site reducing the footprint of the building to nearly half the floor area of the upper level.

Acapulco's hills are made up of huge granite rocks piled on top of each other. In a purely mimetic endeavor, the designers worked hard to make the chapel look like just another colossal boulder atop the mountain.

Formative Arts

077

建筑造型和建筑结构

蓝色森林
Blue Forest

项目档案	Project Facts
设计：Material Landscape	Design: Material Landscape
面积：200 平方米	Site Area: 200 m²

这个庭院位于 Nissan 设计工作室和会议室之间。屋顶并没有全部封闭，而是露出狭长的一块，让阳光和天空的美景可以洒射进来。当人们进入这个庭院的时候，感觉就像是走在一个过道里，过道向远方延伸而去。顶部采用蓝色的树干构成一个华盖，不同的树干，在阳光照射下，呈现出多变的倒影。

Inserted amidst Nissan's design studios and boardrooms, this courtyard offers a long slot of open sky that unfurls like a picturesque parkway vanishing in the distance. The designers reinforced this impression by suspending a canopy of branches overhead, which casts ever-changing shadows on the ground. However, the sensations of travelling under a forested promenade are defied as the gaze turns upward: the branches are cast in fibreglass and painted sky-blue. Blue Forest thus challenges conventional preconceptions of nature, while intensifying the perception of light and colour.

Formative Arts

建筑造型和建筑结构

犹太人被驱逐出境纪念馆
Jewish Deportation Memorial

项目档案

设计：Studio Kuadra
项目地点：意大利

Project Facts

Design：Studio Kuadra
Location：Italy

这个纪念馆位于火车轨道和主要街道之间，人们前来参观都很方便。

纪念馆的表面是由混凝土铺设而成的，并且混凝土路面略高于地面，看去来就像是火车站用于货车停靠的平台。这个平台的周围是不同大小的石头。20个幸存者的名字的拼写立在平台上，材料上选取柯尔顿耐腐蚀钢，给这些名字增加了历史厚重感。聚光灯的照射，一方面纪念这些犹太人，同时也温暖了钢材的硬冷。在平台上还有350块瓷片用来纪念那些被驱逐出境的犹太人。一条长长的红色条带从火车站一直延伸到纪念馆入口处的斜坡。在斜坡处有一块信息板，标明了历史。

Formative Arts

The memorial is situated between the train line and the main road that leads to the French border from Cuneo, and highly visible and accessible to the public at all times. The base of the memorial is a concrete slab that has been slightly raised off the ground as if it was the platform for the freight wagons. The platform is surrounded by rocks of different sizes. On the platform, the names of the 20 survivors are spelt out in three-dimensional letters in corten steel, while on the ground, 350 plaques commemorate the deportees that the freight wagons, caused by oxidation. The memorial is lit up by spot lights positioned at the base of the name pillar of each survivor, while a series of hidden lights give the illusion that the base is lightly raised with respect to the ground. A long red strip on the pavement leads from the adjoining train station to the access ramp at the memorial where there is a plaque explaining its purpose and the history of what happened there.

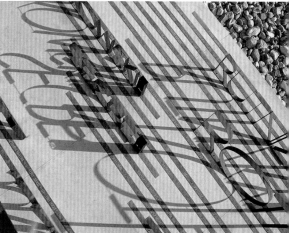

建筑造型和建筑结构

Crater Lake
Crater Lake

项目档案

设计：24° Studio (Fumio Hirakawa + Marina Topunova)
项目地点：日本，神户
尺寸：直径 10 米，最高处 1.7 米
完成时间：2011

Project Facts

Design：24° Studio (Fumio Hirakawa + Marina Topunova)
Location：Kobe, Japan
Dimensions：10m DIAMETER x 1.7 m HEIGHT (MAX)
Year：2011

设计的灵感来源于 1995 年阪神的大地震导致建筑环境不可避免被损坏，这种破坏性的历史影响让神户的居民变得更加坚强和团结。亲密的社会关系能帮助他们战胜灾害重建城市，使其成为更好的生活环境，让人与人的社会关系变得更为密切。在 Crater Lake 景观设计之初，不仅考虑满足周围环境的融洽，更为重要的是强调社会之间的互动。Crater Lake 外表像是美国奥瑞根的漪丽火山湖，坐落于神户的人造岛上的 Shiosai 公园，为神户的市中心提供了一个宏大的山景和海景图。考虑到该地点地理位置的特殊性和优势，设计中选用木质材料搭造一个起伏景观，提供了一个开放不受约束且能 360 度观景的视觉景观。景观中的每一个表面都适合人们坐下来或是躺下来，在空间的中间也有一些座椅。

卓越的设计想法和合适的材料选择创造了这个光滑耐用的复杂综合景观。之所以选择木制材料，不仅是因为结构上的可塑造性，也是因为天然这一特点。光滑的连续的木制表面并没有经过技术处理，而是将圆形的表面分割成 20 个放射状的部分。当然，这 20 个部分不是固定的，需要根据具体的情况来决定。这 20 个部分都是预先安装好，然后运到这个公园的。

放射状部分由形式不统一的 64 个木片构成，这些木片在水平方向上和垂直方向上互相支撑。在人流较多的部分，木片和木片之间有狭窄的缝隙。越往上，木片之间的空隙越大。这样方便人们向上攀爬。这样的设计来源于当地的地形和四季特点，同时也考虑到功能上的要求——既可以遮阳也可以避风。

Formative Arts

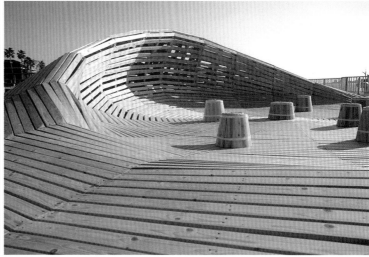

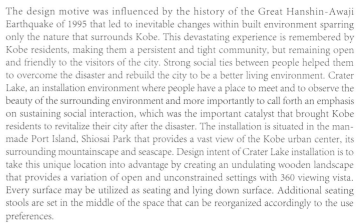
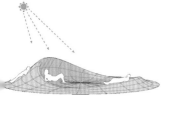

The design motive was influenced by the history of the Great Hanshin-Awaji Earthquake of 1995 that led to inevitable changes within built environment sparring only the nature that surrounds Kobe. This devastating experience is remembered by Kobe residents, making them a persistent and tight community, but remaining open and friendly to the visitors of the city. Strong social ties between people helped them to overcome the disaster and rebuild the city to be a better living environment. Crater Lake, an installation environment where people have a place to meet and to observe the beauty of the surrounding environment and more importantly to call forth an emphasis on sustaining social interaction, which was the important catalyst that brought Kobe residents to revitalize their city after the disaster. The installation is situated in the man-made Port Island, Shiosai Park that provides a vast view of the Kobe urban center, its surrounding mountainscape and seascape. Design intent of Crater Lake installation is to take this unique location into advantage by creating an undulating wooden landscape that provides a variation of open and unconstrained settings with 360 viewing vista. Every surface may be utilized as seating and lying down surface. Additional seating stools are set in the middle of the space that can be reorganized accordingly to the use preferences.

Multiple ideas and materials were tested to realize the complexity of smooth and undulating form. Wood was chosen for its strong structural capacity, ease of work with, and natural qualities. One of the main issues was to express continuous and smooth surface without using costly techniques of wood steaming, bending or digital fabrication. The solution was to divide the circular surface into a number of radial parts, with optimal number of 20 parts. Factors that determined this optimal number were overall surface expression, production schedule, and transportation method (vehicle bedsize). These 20 radial parts were preassembled off the site and transported by a vehicle to the main site of Shiosai Park.

The structure of radial parts consists of series of free-form ribs composed in segmentations with horizontal support and cross bracing for rigidity. Each radial segment has 64 surface planks that are attached to three structural ribs rigidly connected between each other with horizon-tal supports. The surfaces with the most anticipated traffic flow have narrow spacing between each plank. And as the mount becomes higher, the spacing distance of surface planks increases, allowing users to climb the mount. The rising mount resulted from understanding the site and seasonal conditions functions as a sun shading and wind protection from the bay winds.

建筑造型和建筑结构

马丁路德教堂
Martin Luther Church in Hainburg

项目档案	Project Facts
设计：Coop Himmelb(l)au	Design：Coop Himmelb(l)au
项目地点：澳大利亚，海恩堡	Location：Hainburg, Austria
面积：420 平方米	Site Area：420 m^2
完成时间：2011	Year：2011

建筑的形状宛若一个巨大的钟表，屋顶的"表"依靠四个钢柱支撑。另一个重要的语言是祈祷室的上方，其弧形屋顶的设计语言源自罗马风格的藏尸骨罐子，设计师将此语言用现代数字工具转换成一个符和时代的造型。这个项目的特殊的地方也在于对光及透明度的发挥。扭曲屋顶的三个巨大天窗将光线引入室内。三个天窗的数量三，与基督教的三位一体概念契合，可以理解为一个"蓄意的巧合"。教堂本身不仅仅是一个神秘和宁静的地方，在这个快节奏的媒体时代，它也具备了社会开放空间的功能。

The shape of the building is derived from that of a huge "table" with its entire roof construction resting on the legs of the "table"——four steel columns. Another key element is the ceiling of the prayer room: its design language has been developed from the shape of the curved roof of a neighboring Romanesque ossuary, the geometry of this century, old building is translated into a form, in line with the times, via todays digital instruments. The play with light and transparency has a special place in this project. The light comes from above: three large winding openings in the roof guide it into the interior. The correlation of the number Three to the concept of Trinity in the Christian theology can be interpreted as a "deliberate coincidence".

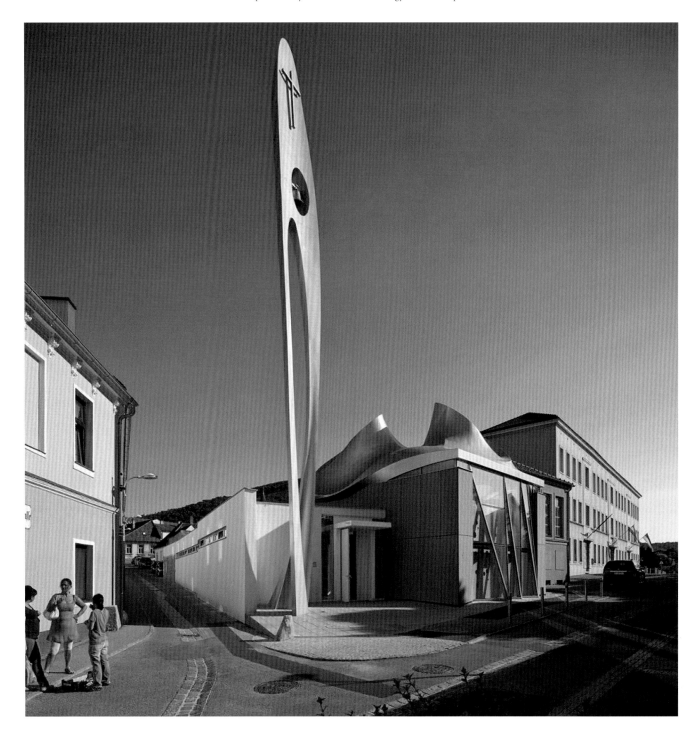

口红森林
Lipstick Forest

项目档案	Project Facts
面积：700 平方米	Site Area：700 m²
项目地点：加拿大，蒙特利尔	Location：Montreal, Canada

这个项目打破惯例，没有利用盆栽植物去抵御当地的气候，而是栽植了 52 棵用混凝土构成的树木，树木被涂上粉色的口红颜色，不仅代表了这个城市日益繁荣的化妆品工业，而且聚集了蒙特利尔无限的生活乐趣。地面的设计类似篮球场地板，不仅历史悠久更具有说服力，而且适应了未来的城市中心环境。

Instead of trucking in potted plants and fighting against the local climate to keep them alive, the designers planted a forest of fifty-two concrete trees, painted lipstick-pink to celebrate the city's flourishing cosmetic industry and manifest Montreal's inexhaustible joie de vivre. Patterned after the hundred-year-old maples that line the avenues in the old city, the forest is perfectly adapted to the futuristic environment of the city center.

建筑造型和建筑结构

O-STRIP 展厅
O-STRIP Pavilion

项目档案

设计：Shuang Gao, Jing Shao, Shengkan Zhang
项目地点：中国，上海，同济大学
完成时间：2011

Project Facts

Design：Shuang Gao, Jing Shao, Shengkan Zhang
Location：Tongji University, Shanghai, China
Year：2011

O-STRIP 展厅是同济大学制造工作室设计的一个项目，它的目的是为所在场地带来生机。这里是教学楼附近的一个下沉空间，它与一个室内展厅相连，在非展出时间，这片空间基本上闲置，无人问津。为了改变这一现状，设计师认为关键是新结构的强大的代表作用，起到吸引路人的效果。电脑技术使复杂的结构变得清晰，同时还加速了制造过程，这个展厅既是电脑技术的表现，同时它还能为学校不同的活动提供服务。

O-STRIP Pavilion is one of the projects in Tongji University's Fabrication workshop that aims to bring back the liveliness of the place. The site is located at a sinking space of the teaching building, which connects with an indoor exhibition hall. In non-exhibition time, there is few people using this space, and the space itself becomes negative. How to energize this place? The key is the strong representational power of the new structure, which can attract passers-by. Computational technology makes complex-structure clear and speeds up fabrication process. At the same time, the pavilion will be available for all kinds of campus activities.

Formative Arts

建筑造型和建筑结构

ONE Kearny 大厅
ONE Kearny Lobby

项目档案

设计：IwamotoScott
项目地点：美国，洛杉矶
完成时间：2010

Project Facts

Design: IwamotoScott
Location: San Francisco, USA
Year: 2010

这个大厅是为洛杉矶一个新的商业区设计的，大厅的主楼为一座新建的12层大楼。一系列钻石状的透明木制花格镶板构成了大厅的主要入口，看起来像是一个发光的天花。为了映衬花格镶板的有脚表面，一系列小面积的表面从接待台延伸开来，一直到锥形的走廊尽头。

Designed for a new downtown San Francisco development at the convergence of Market, this lobby serves a new twelve storey annex. A series of diamond-shaped translucent wood coffers define the main lobby entrance, acting as a lighted modular ceiling. Projecting the language of coffers' angled surfaces, a series of faceted surfaces move from the reception desk to the walls that lead to the elevators at the end of a tapered and inflected hallway.

建筑造型和建筑结构

环形穹顶
Ring Dome

项目档案　　　　Project Facts

设计：Minsuk Cho　　　Design：Minsuk Cho
项目地点：纽约　　　　Location: New York

环形穹顶最初是为纽约艺术与建筑展览馆 25 周年纪念设计的用作店面的临时建筑。这个穹顶形的结构直径为 8.65m，由大约 1 500 个塑料呼啦圈组成。这些呼啦圈可以自由叠加，用大约 12 000 多根拉链绳固定连接在一起。

The temporary pavilion, designed by Korean architect Minsuk Gho of Mass Studies and built of hula hoops, is a new version of the one constructed in New York to celebrate Storefront's 25th anniversary. The pavilion, built out of 1,500 hula-hoops and 12,000 zip-ties, is installed in the Galleria Wtorio Emanuele, the shopping arcade that connects Piazza Duomo and Piazza La Scala.

Formative Arts

 银色公园码头办公楼
Silver Park Quay

项目档案   Project Facts

设计：Wiel Arets   Design: Wiel Arets
项目地点：荷兰   Location: Netherlands
完成时间：2010   Year: 2010

该办公楼从它所在的位置获得了一个颇具浪漫色彩的名字——银色公园码头，它位于 Lelystad 的 Zilverparkkade。所有四个门面的表面都全部或部分覆盖着预制混凝土构件，象征着一个分支状结构。巨型花丝是来自对分形算法的研究。荷兰绘画艺术家 Maurits Escher 确立了模式。这个重复的模式需要特殊技能去制定，用有限数量的的混凝土元素去构成无缝的花纹，同时，避免让人一眼看出花纹是重复的，这种要求与纺织品和墙纸的设计轮换模式相类似。

The project derives its romantic name (Silver Park Quay) from its location: the office cluster on the Zilverparkkade in Lelystad. All four facade surfaces in the design are either entirely or partly covered with prefabricated concrete elements, symbolizing a branch-like structure. This blown-up filigree is the result of a study of infinite patterns. The works of the Dutch graphic artist Maurits Escher have been an unmistakable model. Specific skills are required to devise a repetitive pattern that, applied in a limited number of different concrete elements, constitutes a seamless entity. The craftsmanship needed to avoid the repeating units to be too obvious, has an analogy with designing rotation press patterns for textile and wallpaper.

巨大的波浪形木材装置
Gigantic Timber Wave Installation

项目档案	Project Facts
设计：Helen Morgan	Design: Helen Morgan
项目地点：伦敦	Location: London

这个巨大的波浪形木材装置在维多利亚和艾伯特博物馆的入口展示。优雅的波浪形状给伦敦设计节注入了新鲜的血液，吸引更多的人来到这里。这个木材装置有三层楼那么高，人们可以在装置的下面通过。螺旋状的格子设计是数位建筑师和工程师共同努力的结晶。这个装置不仅呼应而且提升了原来的入口。

This majestic Timber Wave Installation was unveiled at the entrance to the Victoria and Albert Museum, where its elegant undulating form welcomed visitors to the London Design Festival. Transforming the front of the magnificent museum, the three-story-tall timber sculpture towers are giant enough for people to pass beneath it. This spiraling latticework is the collaborative creation of award winning architects along with engineering firm Arup, who responded to the original V&A entrance with a design that enhances the existing grand facade.

Formative Arts

 ## 空灵的钟声挂桥
Ethereal Chimes Hung From Bridge

项目档案	Project Facts
设计：Mark Nixon	Design：Mark Nixon
项目地点：丹麦，奥尔胡斯	Location：Aarhus, Denmark
完成时间：2011	Year：2011

本案是由600根金色的氧化铝管构成的，铝管的直径为50毫米，长度从120毫米到3 450毫米等长度不一。铝管依附在桥的下面，而在桥面上有一系列的交错节点，供游人玩耍。设计想法基于三点：音乐、互动、听觉。当风吹过，这些铝管就引起游人的注意，散发出一种动态之美；当没有风的时候，这些铝管就像完全消失了，不被游人所知。

The piece is constructed from 600 gold anodized aluminium pipes of 50mm diameter ranging in length from 120mm up to 3,750mm. These pipes are attached to the underside of a bridge and with a series of interactive nodes on the top surface that allow for people to play the instrument.

The design is based on three conceptual ideas: the idea of music and interaction as a catalyst for conversation and play, the non-visual object. A constant varying in wind conditions on the site mean that the sculpture will hide and reveal itself through the creation of sound when the wind chooses to blow. Some days the sculpture will be discovered, creating a beautiful moment of realization in the viewer, while other days the sculpture will remain still and may be completely passed by.

城市雕塑
Urban Sculpture

项目档案　　　　　　Project Facts

设计：Helen Morgan　　Design: Helen Morgan
项目地点：斯洛文尼亚　Location: Slovenia

帝沃力公园是一个受到特殊保护的地区，不允许再增建任何建筑，所以设计师们不得不在公园的附近找合适的场地。最终，这个城市雕塑位于一个凌乱的、长满草的三角形地带，正好是在公园入口的旁边。这个雕塑，不仅赋予了草地新的生命力，而且成为公园的新入口。雕塑的颜色鲜艳，而且与周围的环境形成鲜明的对比，结构上更是充满动态之美，带动了整个环境的活力。另一方面，也吸引人们来到这里游玩。

The Park Tivoli is a very protected area, where you are not allowed to build, so the designer had to search for a new location within direct vicinity of the park. The final resting place for this sculpture turned out to be an untidily grassy triangle, located directly by the park entrance. By placing the sculpture here, the designer had revitalised this meadow and established a new entrance to Tivoli park. With a bright and contrasting color, the sculpture is an open, clear and structurally dynamic form that explores the possibility of motion in the environment, while this movement itself is present as a natural environmental process. With its form, this dynamic sculpture stimulates the interest of pedestrians and invites them to pass through.

Formative Arts

建筑造型和建筑结构

本迪戈的中国区
The Bendigo Chinese Precinct

项目档案

设计：Mothers Art (Sculpture Detail Design and Fabrication)
项目地点：澳大利亚，维多利亚州

Project Facts

Design: Mothers Art (Sculpture Detail Design and Fabrication)
Location: Victoria, Australia

本迪戈的中国区位于本迪戈小湾，这里是城市中心的边缘，旁边还有植物园。新的中国区面积很大，能够举办各种大型活动。整个空间有大量美化的露台，人们进入和休息都很方便；而且被分为两个层次。低层的广场可以作为一个圆形竞技场的前场，地面是用中国的花岗岩铺设的，另外还有河流的三角洲式样作底，并刻有本迪戈的中国式名字——Big Gold Mountain。空间的上层连接着河流，是由当地的石板铺设而成的，这种石板颜色纯天然，并有黄铁矿斑点。植物的选取上遵从中国的历史和景观，松柏和竹子是首选，而在低层则选用了当地的物种和沼泽地植物。

The Bendigo Chinese Precinct is located on the Bendigo Creek at the edge of the city centre and botanic parklands. The new precinct provides a much needed large open air plaza to host public events. The space is split over two levels with landscaped terraces providing seating and access. The lower plaza acts as an amphitheatre forecourt and is paved in Chinese granite with the river delta pattern referencing the homeland of the Chinese miners and is inscribed with the historic Chinese name for Bendigo——Big Gold Mountain. The upper level bridges the creek and is paved in a local slate with natural color variation and iron pyrites speckling.
Planting is inspired by Chinese history and landscapes, with conifers and bamboo forming the primary planting character. Low level indigenous species and swale planting are part of the water sensitive urban design.

Formative Arts

建筑造型和建筑结构

夏季展厅
Summer Pavilion

项目档案

设计：Danecia Sibingo, Lyn Hayek, oojin Kim, Taeyoung Lee
项目地点：伦敦

Project Facts

Design: Danecia Sibingo, Lyn Hayek, oojin Kim, Taeyoung Lee
Location: London

这个展馆位于伦敦的贝德福德广场。设计师们将之称为"浮木"，给人一种极好的感官和空间效果。原形是通过电脑设计出来的，注重线条的流动之美。最终的展馆是由 28 块胶合板构成，系统结构性非常好。

This pavilion is in London's Bedford Square. The designers have named their work "Driftwood" with a sensuous and overwhelming spatial effect. Their ideas were manifested through a computer-generated script which manipulated the movement of lines in a continuous parallel fashion, creating line drawings which formed the basis of a plan. The final design consists of twenty-eight layers of plywood which conceal an overall internal "kerto" (a renewable spruce plywood) structural system.

建筑造型和建筑结构

豪华的帐篷建筑
Regent's Place Pavilion

项目档案	Project Facts
设计：Carmody Groarke	Design: Carmody Groarke
项目地点：英国	Location: Britan

设计师意在创造一个公共的细长的塔器，其顶棚有 8 米之高。顶棚下面密度不一的立柱在白天散发出夺目的光彩，在晚上则散发出金色的光芒，为路人带来了视觉上的莫阿效应。在临街的两端，这个帐篷都可见，并且增加了临街的人流量。顶棚是由不锈钢构成的，成网格状，厚度为 3 毫米，下面的立柱没有任何交叉支撑。这个帐篷在重量，高度以及本身建筑特点上都与临街和周围的环境和谐统一。定做的 LED 灯以鹅卵石的方式呈现，照亮了整个帐篷，而且也为临街提供了照明。

Carmody Groarke's concept for the pavilion presents a pavilion as an open field of slender columns which supports a canopy eight metres above the landscape of the street. Visible from Euston Road, the pavilion reveals various clustered densities of the vertical columns beneath its canopy, which shimmers in sunlight by day and contain intense projected "gold" light by night, generating a visual moires effect for passers-by. Its dramatic form is visible from approaching each end of Triton Street intensifying the experience of movement between 10 and 20 Triton Street. Holding the 3mm plate stainless steel canopy, extremely slender vertical elements stand without any cross-bracing, joined only at the top with a decorative structural lattice.

Formative Arts

The pavilion forms a lightweight counterpoint to the architecture of the public colonnades flanking each side of the street, relating architecturally to the height of these adjacent but also inviting views across the street from one side to the other. Amongst the field of elements, bespoke LED lighting is set into the pattern of the cobbled surface to up-light the pavilions canopy, providing all the ambient external lighting to this end of Regent's Place.

建筑造型和建筑结构

空间灵魂展馆
Movie by Spirit of Space

项目档案　　　　　　　Project Facts

设计：Studio Gang　　　Design: Studio Gang
项目地点：美国，芝加哥　Location: Chicago, USA

这个展馆是由弯曲的木材构成的，整体成晶体结构。表面上采用的材料是玻璃丝。不管白天还是晚上，这里都能为人们提供便利。设计师们还在一个19世纪的池塘周围设计了木板路，改善了水质以及当地野生物种的生活环境，同时也为人们提供了学习的好地方。通过改善场地的水文、景观、可达性，这个展馆可以被用作户外教室。整个展馆的设计灵感来源于一种龟甲的壳，片状结构是由预制的弯曲木块以及一系列的内部相连的玻璃丝构成的。

The movie shows the construction of the bent-wood lattice structure, the installation of fibreglass shells to the top of the pavilion providing shelter and its various uses throughout day and night. Studio Gang Architects also designed the boardwalk surrounding the 19th century pond, improving the water quality and habitat for the local wildlife and creating an educational nature trail. With the design's improvements to water quality, hydrology, landscape, accessibility, and shelter, the site is able to function as an outdoor classroom in which the coexistence of natural and urban surroundings is demonstrated. Inspired by the tortoise shell, its laminated structure consists of prefabricated, bent-wood members and a series of interconnected fiberglass pods that give global curvature to the surface.

Formative Arts

 大运河广场
Grand Canal Square

项目档案	Project Facts
设计：Martha Schwartz	Design: Martha Schwartz
项目地点：爱尔兰，都柏林	Location: Dublin, Ireland

这个项目最大的设计特点就是"红地毯"，这个"红地毯"从歌剧院一直延伸到码头。另外还有一个"绿地毯"与之交叉，形成强烈的对比。"红地毯"是由树脂玻璃铺设而成的，表面上覆盖有红色的方块，每个方块的形状都不一样。"绿地毯"连接着新酒店和办公楼，绿化带构成不同的多边形，为人们提供了舒服的休息场所。绿化带的植物多是沼泽类，不仅柔和了整个空间，而且会让人们想起这里原来是湿地。

The design features a striking composition of a red carpet. It extends from the theatre both into and over the dock, and is then crossed by a lush "green carpet" of planters with lawns and vegetation. The red carpet is made of bright resin-glass paving, covered with red glowing angled light sticks. The green carpet of polygon-shaped planters offers ample seating, and connects the new hotel to the office development across the square. The planters themselves feature marsh vegetation which softens the space, and acts as a reminder of the historic wetland nature of the site.

建筑造型和建筑结构

投掷
Drop

项目档案　　　　　**Project Facts**

设计：Paul Cocksedge　　Design: Paul Cocksedge
项目地点：伦敦　　　　Location: London

设计师 Paul Cocksedge 在 2010 年伦敦设计节上的展品是一个大大的立着的金色弯曲圆盘。这个作品表面是有磁性的，意味着路人可以将自己多余的硬币粘在圆盘表面。活动结束后，这些收集来的硬币将捐给儿童慈善机构。

The designer Paul Cocksedge has installed a large buckled disc outside the Southbank Centre in London. The piece is magnetic, meaning passers-by can attach their unwanted pennies to its golden surface. Once the installation closes, every penny attached to the disc will be donated to children's charity Barnardo.

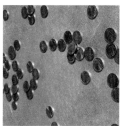

Formative Arts

Hyparform 纵帆
Hyparform Vertical Sails

项目档案	Project Facts
设计：Planex	Design: Planex
项目地点：奥地利，维也纳	Location: Vienna, Austria

维也纳一个公司大楼的立面是由104块帆布构成的，一方面挡住了阳光，另一方面充当了建筑的另外一层表面。Planex公司利用高准确度的仪器测量这些纵帆，并且利用高频焊接仪器来焊接这些帆布。在制造过程中，帆的弯曲度以及纬线都是预先压制好的，这样就保证了较好的稳定性以及抗风性，而且，能够保证在不同的温度之下，帆不至于会变形。

The facade of a company building in Vienna has been fitted with a total of 104 sails which on the one hand protect the building from the sun and on the other act as architectural cladding on the closed side of the building. The firm of Planex calculated the dimensions of each sail by modelling based on highly accurate measurements, and all the seams were welded using high-frequency welding equipment. During manufacturing both the warp and the weft yarn is pre-stressed and then coated. This guarantees a high level of surface stability and is a precondition for ensuring adequate wind resistance, which can only be guaranteed if the sails neither expand nor contract in response to changes in temperature.

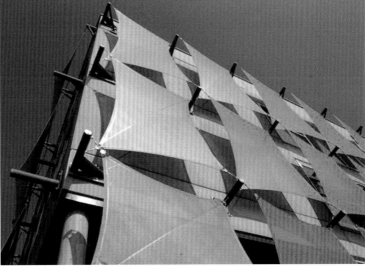

非线性装展馆
NonLin/Lin Pavilion

项目档案

设计：Marc Fornes
项目地点：新奥尔良

Project Facts

Design: Marc Fornes
Location: New Orleans

这个展馆不是一个大结构的模型或者建筑，不是一个艺术装置，也不是由回形针连接的硬纸板。它的结构完整性依赖于一整块材料的完全利用，而且具有防水性。看似轻巧，但是非常稳固。你可以在上面坐着，挂着，甚至是攀爬。这个展馆是一个原型建筑，形态紧密，由多个不同形状的"Y"形结构构成。这种原型结构是一种形式到另外一种形式的转换，在打开的同时，将本身的小结构融入到更大的结构单元中。这样，表面的密度就得到加强。

The Pavilion is not considered a model of a larger structure or a building, neither is it an art installation. It is not made out of cardboard, or connected through paper clips. Its structural integrity does not rely on any camouflaged cables and it can resist water. It is light yet very strong. One could sit on it, even hang or climb it. It is a prototypical architecture. The cohesive morphology of the pavilion originates from a "Y" model referred to as the basic representation and lowest level of multi-directionality. This prototypical structure is an investigation into transformations from one state to the other. Members within the structural network are opening up and recombining themselves into larger apertures while their reverse side is creating a surface condition providing that as density increase eventually provides to the person evolving within a sensation of enclosure.

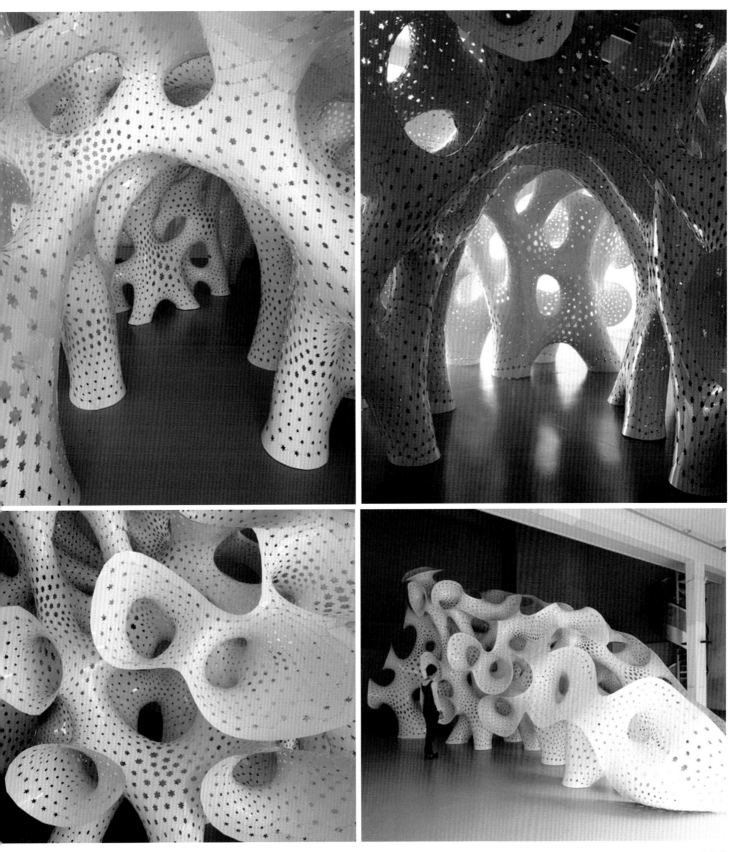

水晶网
Crystal Mesh

项目档案

设计：WOHA Architects
项目地点：新加坡

Project Facts

Design: WOHA Architects
Location: Singapore

水晶网立面总面积为 5 180 平方米，其中 2 550 平方米安装了多媒体装置。所有的水晶都是由透明的聚碳酸酯外壳和铝制板构成的。所有的水晶被分为两类，一部分水晶有 1-7 个可以单独控制的灯源；而另外一部分水晶不带任何光源。

The Crystal Mesh facade covers a total area of 5,180m^2 of which 2,550m^2 are equipped as a media installation. All "Crystal Elements" are made up of a translucent, polycarbonate hull and an aluminum back-plate. Each electrified Crystal Element contains between 1 and 7 individually controllable light sources, which are located in the respective "Light Cells" of the Crystal Element. In contrast, there are "blind" Crystal Elements, which carry no light sources inside.

Formative Arts

墨西哥独立 200 周年纪念 "火炬"
Bicentennial Torch

项目档案　　　　　Project Facts

设计：Jose Pareja　　　Design: Jose Pareja
项目地点：墨西哥　　　Location: Guanajuato, Mexico

纪念雕塑下方 10 米是混凝土材料，上方 35 米是 100 个环状钢构组成的形体，环和环之间又形成 100 个空隙，透过这些环状结构的空隙，可形成丰富的光影效果。夜幕降临，雕塑就会变成城市里一个伟大的灯具，照亮周边环境。从形体缝隙中透出的光芒璀璨迷人。形体上有 200 个印记代表着墨西哥的独立岁月的奋斗，夜间赫赫生辉的 100 道光芒象征着百年革命运动及其对墨西哥的重要性。45 米高处的顶部能向天空投射具有力量的光束，就像是象征民族独立和未来的永生之焰。

The monument consists of a 10-meter tall concrete volume, followed by a 35-meter steel structure made of one hundred rings, interspersed with a hundred voids, which act as optical negatives as a result of the shadow projected onto them. By nightfall, the sculpture becomes a great urban lamp, which aims to enlighten the surrounding environment through its body and scars. Its body, marked by 200 scars, reveals on its skin and its longevity, giving life at night to a hundred rings of light, representing also the centennial of the revolutionary movement and its importance in the independent living of Mexico today. The rings are crowned at 45-meter high with a cauldron from which a powerful beam of light is projected to the infinity, a perpetual flame symbolizing the independence and the nation's future.

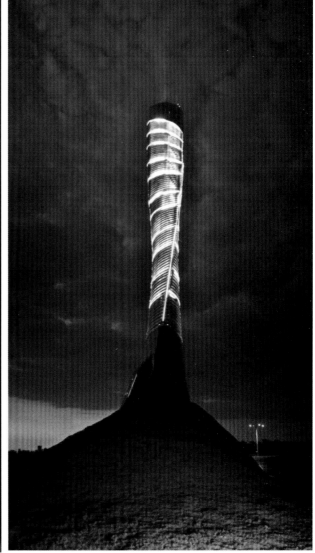

Tubaloon
Tubaloon

项目档案　　　　**Project Facts**

设计：Susanne Fritz　　Design: Susanne Fritz
项目地点：瑞士　　　Location: Zurich, Switzerland

Tubaloon 是一件为哥尼斯堡国际爵士音乐节设计建造的大幅度弯曲、线网可伸长的亭阁状充气建筑。Tubaloon 整个结构看起来就像是人体的器官，而内部的结构看起来就像是骨骼。因为每年这个建筑都会拆卸，所以所选取的材料就是聚氯乙烯薄膜，不仅经久耐用，而且对于光线的投射以及声音上的要求都是非常合适的。

Tubaloon is a pneumatic membrane sculpture which was created for the Kongsberg jazz festival in Norway. The combination of a stretched membrane with a pneumatic design is the special feature of the Tubaloon: stretched membrane structures have exterior structural parts but in the case of the Tubaloon the static structure is located like a skeleton inside the pneumatic shell. This gives the structure a body-like, organic look. Every year the Tubaloon is set up and dismantled once more and so a white, PVC-coated PVC-PES polyester membrane by Ferrari was selected because of its hardwearing properties and suitability for projection and acoustic purposes.

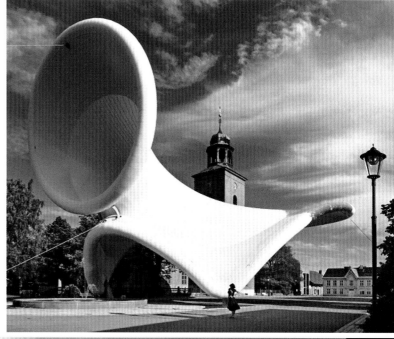

Formative Arts

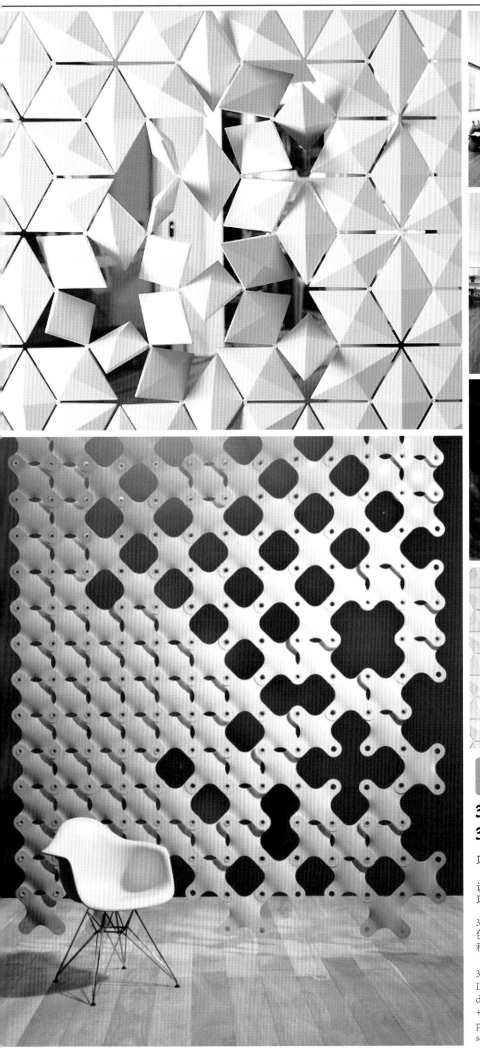

3form 新设计产品
3form New Ntunning Products

项目档案	Project Facts
设计：Susanne Fritz	Design: Susanne Fritz
项目地点：荷兰	Location: Netherlands

3form 是领先的建筑和设计制造商，主要从事天花板装饰，色度，光板，房间隔板等的设计和制造。凭借环保材料和先进的解决方案，3form 最近设计了四个新的产品。

3form, the leading manufacturer of ceiling elements, Chroma, Ditto, light panels, light line, room divider, shapes, space divider architectural hardware solutions for the Architecture + Design industries, has recently launched four new, stunning products: Ditto, shapes and facet plus new a integrated lighting solution——light line.

113

建筑造型和建筑结构

Casalgrande 陶瓷云
"Cloud" Casalgrande Padana

项目档案

设计：Kengo Kuma & Associates
项目地点：意大利

Project Facts

Design: Kengo Kuma & Associates
Location: Italy

该项目位于一块超过 2 800 平方米的公共绿地上，"陶瓷云"第一次运用陶瓷作为结构材料。Casalgrande 公司的大型元件通过标准化的产品摆脱了通常使用的简单涂层并将之转化成一种质感和令人难忘的三维空间光影效果。最终的设计呈现出一个三维空间结构，它运用了创新的陶瓷元件。大块的炻瓷板经机械处理和特殊设计的金属件相连组成了这些元件。作为整体，这个装置长 40 米，高 7 米，看起来就像一座巨大的建筑体。

The project is located in one public green of more than 2,800 square meters. For the first time, "Cloud" Casalgrande Padana uses ceramics as its structural materials. Casalgrande gets rid of the usual and simple coating but creates an unforgettable three-dimensional space through standardized products. The final design shows a three-dimensional structure with the use of the innovative ceramic components. The large stoneware plates are connected together and composed of these elements by mechanical treatment and the special design of the metal member. As a whole, this device is 40-meter long, 7-meter high, and looks like a huge construction body.

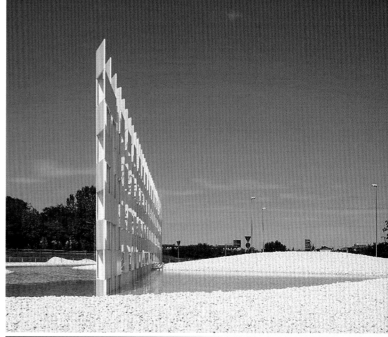

Formative Arts

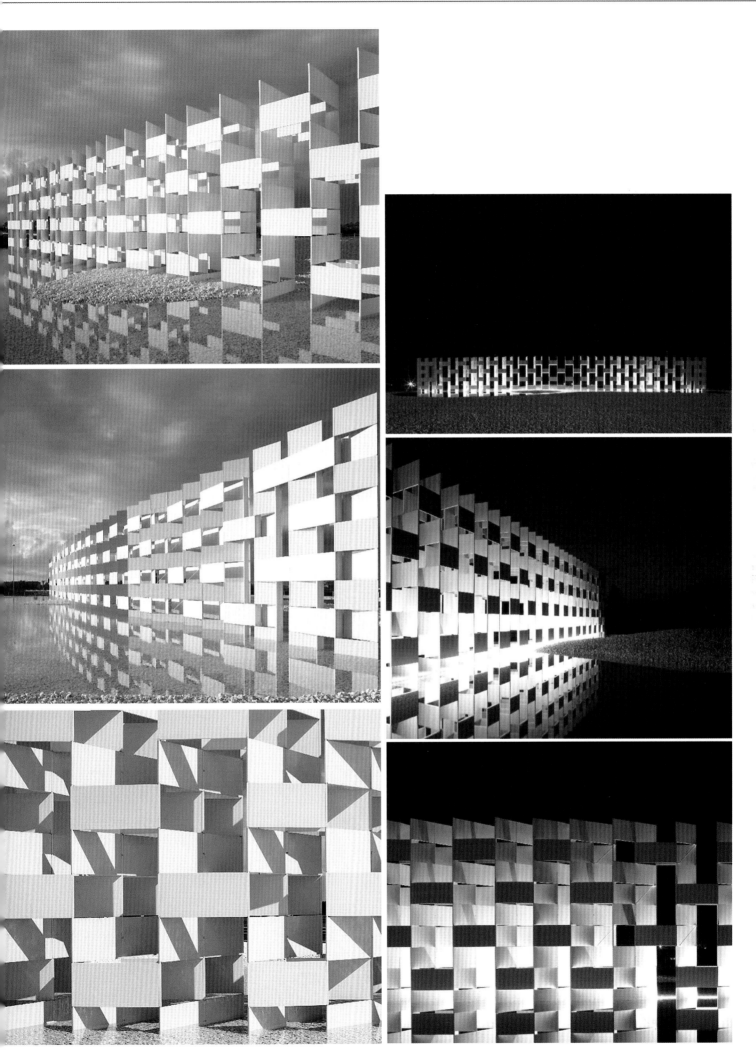

115

3013 装置艺术
3013 Installation

项目档案

设计：Lawrence Lek, Onur Ozkaya, Jesse Randzio
项目地点：伦敦

Project Facts

Design: Lawrence Lek, Onur Ozkaya, Jesse Randzio
Location: London

伦敦建筑协会的学生建造了一个叶子状的装置。这个装置从一个四楼的屋顶平台蜿蜒而下，直到一楼的庭院。三个独立分开的部分形状各异，都是用回收的展示板经过两两扭曲在一起形成的。每两块展示板的边缘扭曲在一起，形成一个非常灵活的表面，并且与周围的环境相得益彰。装置轻柔地从屋顶蜿蜒而下，悄悄地搭在墙壁上。

Students at the Architectural Association in London have constructed leaf-like sculptures that curl down from a fourth-floor roof terrace to a ground level courtyard. Strips of plywood from recycled exhibition panels are twisted into pairs and fastened together using cable-ties to create the three separate parts of the 3013 Installation. They are joined together at their edges to form flexible skins tailored to the site. The upper skins are suspended from above, lightly touching the existing brick walls for support.

Formative Arts

建筑造型和建筑结构

奥迪百年雕塑
Audi Centenary Sculpture

项目档案	Project Facts
设计：Gerry Judah	Design: Gerry Judah
项目地点：英格兰	Location: England

这个重达 44 吨，高达 32 米的雕塑是为了纪念奥迪 100 周年而设计的，亮点就是两辆车冲向云霄。其中一辆车是奥迪于 1937 年推出的一款跑车，而另外一辆是最近推出的"R8 V10"跑车。两辆车的轨道采用七弦竖琴的结构。

The 44 tonne and 32-metre-high sculpture, created to mark the car brand's centenary, features a vintage Audi and a modem car racing into the sky. Designed by Gerry Judah, it celebrates Audi's achievements in motor sport with the legendary 1937 Auto Union Streamliner and the recently launched R8 V10 sports car at either end of a dramatic "swoosh" of lyre tracks, as if they are driving off into the sky.

Formative Arts

建筑造型和建筑结构

CAAC 广场
CAAC / Paredes Pino

项目档案

设计：Paredes Pine
项目地点：西班牙，科多巴

Project Facts

Design: Paredes Pine
Location: Cordoba, Spain

项目设计成一个覆盖区域，不受天气的影响，这将为每周两天的临时市场和其他时间的各类活动提供必要的场所。整个项目由高度和直径都不尽相同的预制圆形构筑物组成，灵活随意的排布方式给人一种似曾相识的城市森林下阴影的感觉。这些太阳伞结构解决了人工照明的同时，也实现了内部排水。

It offers a covered area protected from the weather, which will house a temporary market two days a week and other activities at other times. It therefore poses a solution based on prefabricated circular elements that vary in height and diameter and arranged in a flexible manner to allow a similar vision of an urban forest of shadows. The parasols also solve the artificial lighting in the same item and allow drainage of water inside.

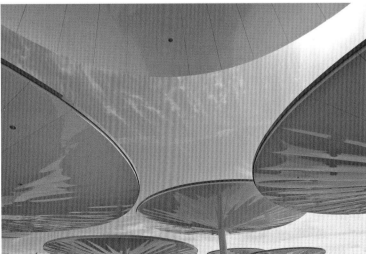

建筑造型和建筑结构

Cirkelbroen 环形桥
Cirkelbroen

项目档案

设计：Olafur Eliasson
项目地点：丹麦，哥本哈根

Project Facts

Design: Olafur Eliasson
Location: Copenhagen, Denmark

国际著名的丹麦冰岛裔艺术家 Olafur Eliasson 在哥本哈根的 Christianshavns 运河上设计了一座环形桥"Cirkelbroen"，为连接 Christianbro 和 Applebys Plads 提供了一条弯曲的步径。2012 年建成后，哥本哈根的市民和游客可以首次在整个内港沿线跑步、行走或骑自行车。
Olafur Eliasson 说："我希望人们把这座桥当作公共广场。我没有设计很长的笔直的码头，而是一个弯曲的桥梁，可以减少速度，转移焦点。我也没有设计一个能最快跨越运河的通道，而是通过小小变化让人们看到城市和公共空间的重新对话。"
根据 Christianshaven 的历史和运河边的文化，Eliasson 采用帆船作为桥梁的视觉起点。环形桥用 5 座错列的大小不一的圆形平台组成，各自建有"桅杆"。

The internationally acclaimed Danish-Icelandic artist, Olafur Eliasson, has designed a new bridge, Cirkelbroen (the circle bridge) which creates a winding path across Christianshavns Kanal in Copenhagen connecting Christiansbro with Applebys Plads, From 2012 when the bridge is completed, the city's residents and visitors is able to run, walk or cycle around Copenhagen's entire inner harbour for the first time. Olafur Eliasson said: "It is my hope that people will stay on the bridge, use the bridge as a public square. In contrast to the long, straight pier, the winding bridge will reduce speed, turn focus. Rather than offering the fastest possible passage across the canal, the bridge will create small variations in the way we see the city and open for a renegotiation of public space." Based on Christianshavn's history and the culture around the canals, Eliasson has used the sailing ship as his visual starting point for the bridge. Cirkelbroen consists of five staggered circular platforms of various sizes each with their own "mast" The bridge is a gift to the Municipality of Copenhagen from the Nordea-fonndation, which is situated in Overgaden neden Vandet right on the canal of Christianshavn.

Formative Arts

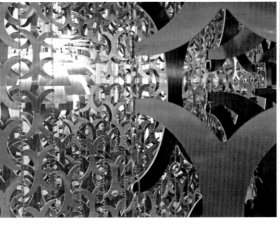

 星云
Nebula / Cecil Balmond

项目档案　　　　　　Project Facts

设计：Belmond Studio　　Design: Belmond Studio
项目地点：意大利，米兰　Location: Milan, Italy

星云由12 000块手工打磨的铝板和4公里长的不锈钢铁链构成。灯光设计在星云的内部和周围，在古典的意境上增添新意，创造了一个情感表达丰富的空间。投影仪的使用，不管白天还是晚上，整个空间都有光与影的不断变化。设计师打破了建筑、科学和艺术之间的界限，创造了一种新的形势。

Nebula comprises over 12,000 hand polished aluminium plates and 4km of stretched stainless steel chain. Lights are positioned within and around Nebula, giving new meanings to classics, creating an emotive and poetic space. Video projections of Balmond's generative forms are projected within Nebula. The result is an ever-changing realm of illumination and illusion, both day and night. Balmond blurs the boundaries between architecture, science and art, creating the generation-next of form.

建筑造型和建筑结构

FibreC 展馆
FibreC

项目档案	Project Facts
设计：Alvin Huang, Alan Dempsey	Design: Alvin Huang, Alan Dempsey
项目地点：伦敦	Location: London

这个项目是一个 100 平方米的临时性展馆。其主要材料有混凝土和 FibreC。FibreC 是奥地利一家公司发明的一种材料，具有高强度、稀薄、灵活、可塑性强等特点，这种材料可以用来构建平整的、弯曲的或其他形状。空间和平面上的器件都是专门定做的，这就使得室内到室外的过渡自然，而且边角的缝隙也会被遮盖住，构成一个光滑的表面。

The project is a temporary 100m² pavilion, made of a concrete skin which consists of "FibreC" elements, a material invented by the Austrian company Rieder. "FibreC" is a high-strength, thin, flexible and mouldable material and can be used for flat, curved and all kinds of special shapes. Formed parts and 2D elements are custom-made to achieve flowing transitions from interior to exterior surfaces and a smooth covering for edges.

建筑造型和建筑结构

教学研究展馆
Research Pavilion

项目档案	Project Facts
设计：ICD & ITKE	Design: ICD & ITKE
项目地点：斯图加特	Location: Stuttgart

这是一个教学研究的临时木材展馆。项目通过计算机设计探讨海胆的骨架，并让其转化为实际的建造，这是一种创新，拓展了仿生学与建筑的结合度。展馆的复杂形态由不同几何形状的极薄（6.5毫米）的胶合板组成。

在建筑设计中融入对生物结构及其空间以及结构材料的全面研究测试。设计中采用模块化的系统，达到高度适应性，并提高了性能。联接的方式是设计中的重要节点。分析海胆，并研究出仿生结构的基本框架。多变和方解石般的表面突起，有利于框架的承载力。

The project is a temporary, bionic research pavilion made of wood at the intersection of teaching and research. The project explores the architectural transfer of biological principles of the sea urchin's plate skeleton morphology by means of novel computer-based design and simulation methods, along with computer-controlled manufacturing methods for its building implementation.

The complex morphology of the pavilion could be built exclusively with extremely thin sheets of plywood (6.5 mm).

The project aims at integrating the performative capacity of biological structures into architectural design and at testing the resulting spatial and structural material-systems in full scale. The focus is set on the development of a modular system which allows a high degree of adaptability and performance due to the geometric differentiation of its plate components and robotically fabricated finger joints. During the analysis of different biological structures, the plate skeleton morphology of the sand dollar, a sub-species of the sea urchin (Echinoidea), became of particular interest and subsequently provided the basic principles of the bionic structure that was realized. The skeletal shell of the sand dollar is a modular system of polygonal plates, which are linked together at the edges by finger-like calcite protrusions.

Formative Arts

建筑造型和建筑结构

粉红色的球
Pink Balls

项目档案

设计：Lightemotion, Impact Production
项目地点：加拿大，蒙特利尔

Project Facts

Design: Lightemotion, Impact Production
Location: Montreal, Canada

170 000 个粉红色的气球高高悬挂在蒙特利尔圣凯瑟琳东街上空，在炎热的夏日为人们提供了一个步行的好去处。这些粉红色的球是用塑料做的，共有三种大小，五种不同深度的粉红色。这些塑料球用线连在一起，横跨在街上，两端固定在树干上，高度不一。悬挂的粉红色球共分为九个不同的部分。每部分的样式都是独一无二的。时而紧密，时而稀疏，街道因此变得充满活力。

170,000 pink balls are suspended high in the air to enliven Montreal's Sainte Catherine Street East and transform for the summer into a pedestrian mall. The plastic balls, in three different sizes and five subtle shades of pink, are strung together with bracing wire, crisscrossing the street and stretching through tree branches at varying heights.
The installation has been deployed in nine sections, each section displaying its own unique pattern. The result is a range of spirited motifs, some dense, others open and airy, all reflecting many moods of the street.

Formative Arts

129

建筑造型和建筑结构

高架通道
Walking along This Elevated Pathway

项目档案

设计：Heike Mutter and Ulrich Genth
项目地点：德国

Project Facts

Design: Heike Mutter and Ulrich Genth
Location: Germany

德国杜伊斯堡（Duisburg）南部的一座雕塑型的过山车向公众开放。这座吸引人眼球的过山车是由 Heike Mutter 和 Ulrich Genth 设计的，它将成为当地的一座地标，反映出这座城市不断变化的进取精神。制作成主题公园里过山车形式的这座步道有着复杂的弯曲折角，但就是因为其流动性，当地才表现出将其重建的愿望。它的名字"老虎与乌龟"也是这个意思，既有速度又有"阻塞"的地方，充满了变化。两位建筑师在谈及这一作品时表示，雕塑般的外形有着不可思议的扭曲，反映出过山车的特质，同时这也是一座跨区域的地标。它看上去似乎一直在成长，而无法被赋予一个确切的定义。如果勇敢地爬上去，站在 45 米高的过山车上能够欣赏到鲁尔地区西部的全景。而过山车的基座占据了 44 米 ×37 米的区域，高度为 21 米。这在德国来说是最大的，也是工程意义上的一个杰作。过山车的扶手上安装了 LED 灯泡，在夜晚勾勒出它的框架，使得夜色也无法淹没它的独特身姿。

Walking along this elevated pathway by German artists Heike Mutter and Ulrich Genth is like being on a roller coaster. The 21-metre-high sculptural walkway is named Tiger and Turtle and is positioned upon a hilltop in Duisburg, Germany.

A staircase winds across the surface of the steel structure, which spirals around itself just like the fairground ride. The dynamic sweeps and curves of the construction inscribe themselves like a signature into the scenery and soar till the height of 21 meters. From a distance the metallic glossy track creates the impression of speed and exceeding acceleration.

The visitor can climb the art work by foot. Although the course describes a closed loop, it is impossible to accomplish it as the looping emerges to be a physical barrier. On top, at the highest point of the sculpture, 45 meters above ground, the visitor is rewarded with an extraordinary view over the landscape of the Western Ruhr. It counters the logic of permanent growth with an absurd-contradictory sculpture that refuses a definite interpretation. With 44 ×37 meters base and 21 meters construction height, the sculpture is not only one of the largest in Germany, but also a masterpiece of engineering.

Formative Arts

建筑造型和建筑结构

Surface Deep 园圃
Surface Deep

项目档案

设计：Leire Asensio Villoria, David Syn Chee Mah
项目地点：加拿大，魁北克

Project Facts

Design: Leire Asensio Villoria, David Syn Chee Mah
Location: Quebec, Canada

Surface Deep 是为加拿大魁北克 Metis 国际园艺庆典的天堂花园参观者，在近期依照进入顺序设计并安置的一个新园圃。

园墙是一个始终展现园区历史的元素，入口处的园壁依循组合式的设计并利用建模技术改造为扭动的缎带式外观。波浪状外形是对新的入园顺序的一种呼应及展示，调整入园的行列队伍并同时在其表层植入一个实验性的苔藓园圃。曲面在作为墙面、背景及遮蔽处的功能及联想意象中来回跳跃，同时也为苔藓园圃创造了多面朝向以及不同的微气候。曲面的多重朝向为苔藓提供从接受日光的斜坡到长期阴暗的悬臂结构等多种不同的生长环境。这些不同的微气候使得不同的苔藓分布需符合其特定条件，其中前 11 个单元格植满砂藓，第 12 个单元格种上草藓，而剩下的其他单元格 (第 13 至 22 个) 混合种植着草藓及其他喜阴森林植物，如赤茎藓、毛梳藓等等。

Surface Deep is a new garden recently installed within the entry sequence for the visitors to the Reford Gardens' Metis International Garden Festival in Quebec, Canada.
Revisiting the garden wall, an element that has been a consistent expressive element within the history of gardening, the entry wall is transformed to form a twisted ribbon-like surface with the help of associative design and modeling techniques. Its undulating form is a response to and gesture for a new entry sequence, framing the entry procession while also embedding an experimental moss garden within its surface. The surface flips in function and association between a wall, a ground and a cover while creating multiple orientations and different microclimates for the moss garden. The surface's multiple orientations offers a number of different growing environments for the moss, from slopes exposed to sunlight to constantly shaded overhangs. These microclimates informed the distribution of a number of moss species specific to each condition, where the first 11 units were made with Niphotrichum canescens (a sun-loving species), unit 12 is planted with CaUicladiiim haldanianum while the other units remaining (13 to 22) were made with a mixture of Callicladium haldanianum and other shade-loving, forest species such as Pleurozium schreberii, Ptilium crista-castrensis and others.

建筑造型和建筑结构

云
The Cloud

项目档案	Project Facts
设计：Nadim Karam	Design: Nadim Karam
项目地点：迪拜	Location: Dubai

"云"是在迪拜上空300米设计的一座冒险性旅游城市，仅用象征雨水的倾斜的支架支撑。

设计灵感来源于游牧民。他们的生活与太阳，水和沙子紧密联系，他们的迁移随着无边无际的云而改变。它建造在半空中，与整个地区遍布的摩天楼的总高度形成对比。总面积为20 000平方米，包括一个湖泊、花园、旋转桥、螺旋通道、露台、礼堂和空中运动平台。"云"的入口为一处平坦的空地，并且有一个游泳池，像雨水一样倾斜的支架如同森林一样反射在游泳池水面上，游泳池上方是一个巨大的悬浮的岛屿。

The Cloud is a speculative design for a resort city elevated 300 metres in the air above Dubai and supported on slanting legs resembling rain.

It is inspired by the nomads, whose lives were defined by the rigours of their relation to sun, water and sand, and whose travels followed the borderless movement of clouds. It is a horizontal presence on an elevated platform, an antithesis to the sum of skyscrapers spreading over the entire region. The Cloud is a 20,000 m² landscape, comprising a lake, gardens, rotating bridges, spiraling walkways and terraces, an auditorium and sky-sports platform. The Cloud is approached on ground level from an esplanade with a pool reflecting a forest of inclined columns reaching up to the huge, translucent floating island.

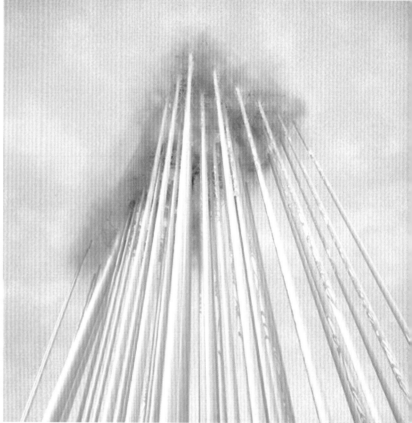

Formative Arts

纯白 Ecoresin
Pure White Ecoresin

项目档案	Project Facts
设计：Nora Schmidt	Design: Nora Schmidt
项目地点：美国	Location: USA

这个项目是对美国马萨诸塞州一栋落后的办公大楼相邻的两个大厅进行整修。建筑师们在没有成本和时间要求的情况下，决定对其进行全面整修。对于毫无生气的外露的电梯，建筑师们提出了一个动态的解决方案，隐藏原来的玻璃窗格。
对于第一个大厅，建筑师们放弃了面板垂直分层的结构，而是在空间和高度上都加以创造性的设计和控制，并选用中性的颜色，这些元素随着电梯的上下，创造出意想不到的效果。对于第二个大厅，虽然仍然选择垂直铁格子结构，但是在立面上更倾向于柔和和飘渺。纯白色的薄板如波浪般的船帆，极具动感之美。

Restoring two adjacent lobbies of the outdated 1980s Adam Place office building in Massachusetts, USA, the internationally operating architects of Perkins + Will wanted to achieve a powerful transformation without the cost and time required for a complete overhaul. Focusing on the unattractive exposed elevators, the team at Perkins + Will chose to update them with a visually dynamic solution for concealing the outdated glass cabs.
For 1 Adams Place, the team suspended 3form Varia Ecoresin panels in a vertical layered pattern. The spacing and depth was controlled to create, along with the nuanced neutral color choices, a sculptural cascade of color and form enhanced by the unexpected elevator movement from within. For 2 Adams Place, the team built on the concept of the vertical grill, but chose a softer, more ethereal approach to the facade. The design called for 3form Varia Ecoresin in brilliant Pure White to emulate billowing ship sails.

未来之塔
Pylon for the Future Competition

项目档案	Project Facts
设计：Ian Ritchie Architects	Design: Ian Ritchie Architects
项目地点：冰岛	Location: Iceland

随着新一代发电站的面世，在未来的几十年里，新的输电线路需求将会加快。整个国家的高压输电线网将会考虑新的输电线路的视觉影响，希望最大程度地与英国乡村景色相融，同时，减少用电量。英国皇家建筑师学会和英国高压输电线网联合发起了一个设计高压线铁塔的比赛。在设计这座塔的同时，比赛的主要目的就是协调能源设施与周围环境之间的关系。

With a new generation of power stations due to come online, in the coming decades, new transmission lines will be needed to carry this new energy to homes and businesses. National Grid will consider the visual impact of its new electricity lines with greater sensitivity to the beautiful British countryside, while balancing this with the need to minimize household energy bills.
The Royal Institute of British Architects (RIBA) for the Department of Energy and Climate Change (DEGC) and National Grid called for designs for a new generation of electricity pylon. As well as exploring the design of the pylon itself, the competition aims to explore the relationship between energy infrastructure and the environment within which it needs to be located.

Formative Arts

建筑造型和建筑结构

C78 枝形吊灯
C78 of Chandelier

项目档案

设计：Mack Scogin Merrill Elam
项目地点：美国，乔治亚州，亚特兰大

Project Facts

Design: Mack Scogin Merrill Elam
Location: Atlanta, Georgia, USA

C78 这个项目被贴切描述为一个被打乱的矢量阵列。这个阵列包括 14 个群集，293 个矢量，29 个魔法球，586 个空间坐标系，1 个圆孔，615 个系链，783 个数字转换器。C78 枝形吊灯和整个大厅空间只是由 52 个嵌入式的灯提供照明。灯散乱地照射在涂漆钢矢量上。手工制作的玻璃魔法球紧紧抓住灯光，为枝形吊灯增添极美的光芒。初晨的阳光洒入大厅，随着时间的迁移，赋予了矢量和魔法球的动态和生命之美。这个枝形吊灯毫无对称性可言，随意而友好，是对现代枝形吊灯的一种新的诠释。

C78 can best be described as a vector array interrupted in its trajectory. C78 consists of 14 clusters of 293 vectors, 29 orbs, 586 space coordinates, 1 oculus, 615 tethers and 783 digital transformations. C78 and the lobby space are simply lit by 52 recessed down-lights. Light plays randomly along the lengths of the painted steel vectors. The hand-blown glass orbs capture light from the same down-lights adding sparkle to the chandelier. In the early morning, east light streams into the lobby, activating the vectors and orbs in a continuously changing kinetic mode. Asymmetrical, informal and friendly, C78 is a contemporary reinterpretation of chandelier.

Formative Arts

100个稻草人军队
Army of 100 Scarecrows

项目档案	Project Facts
设计：Luzinterruptus	Design: Luzinterruptus
项目地点：德国	Location: Germany

100个穿着发光净化服的稻草人组成军队，保卫着汉堡的Dockville节日。稻草人排列整齐，身上有不同的核符号，另外空白的脸和用胶带封住的嘴给人留下深刻的印象。这个项目旨在以一种幽默的方式展示我们正在饱受日本放射性材料的伤害，进而质疑核植物的安全性。

An army of 100 scarecrows dressed in glowing decontamination suits kept a sinister vigil over the Dockville Festival in Hamburg. Installed by Spanish designers Luzinterruptus, the figures were supported in regular rows and adorned with nuclear symbols, blank faces and taped-up mouths. The installation was created for the Dockville Festival in Hamburg which tried to demonstrate, in a humorous tone, the paranoia that we are suffering from since the escape of radioactive material in Japan, has brought into question the safety systems at the nuclear power plants.

Formative Arts

建筑造型和建筑结构

 红球
Red Ball

项目档案	Project Facts
设计：Kurt Pershcke	Design: Kurt Pershcke
项目地点：美国，纽约	Location: New York, USA

红球高度约为 4.6 米，重达 113 千克，是由充气的乙烯基构成的。红球就像是一幅画一样贴在全球各个地方：街上、小巷、桥梁和拱门之间、公交车站。

The Red Ball stands 15 feet high, weighs 250 pounds and is constructed of inflated vinyl. This is not a typical red ball. The Red Ball has been pictured everywhere: on the pavement, squeezed into a tiny laneway, stuck between the supports of a bridge, arch, at a bus stop, at various cities all over the world——the Red Ball is one well travelled piece of plastic!

142

Formative Arts

一阵狂风
A Gust of Wind

项目档案	Project Facts
设计：Paul Cocksedge	Design: Paul Cocksedge
项目地点：伦敦	Location: London

300块婀娜多姿的可丽耐人造大理石经过设计师独出心裁的想法，看起来就像是一堆纸被一阵狂风吹到空中，在空中形成一个纸盘，吸引其他的纸张聚集在这里。每一块大理石都是由设计师亲手打造的。

Three hundred curvaceous pieces of Corian represent a stack of paper that has been blown into the air by a gust of wind. Each of these limited edition pieces is engraved and then handmade by Paul Cocksedge. They will function as paper trays, becoming a place for wandering paper to gather.

巨大的数字折纸老虎
The Giant Digital Origami Tigers

项目档案

设计：Bosse Chris, Rieck Alexander
项目地点：悉尼

Project Facts

Design: Bosse Chris, Rieck Alexander
Location: Sydney

这些巨大的数字折纸老虎采用中国传统的灯笼制作工艺，融传统与创新为一体，将东西方文化结合在一起。老虎均成匍匐状，高2.5米，长7米，只有200千克。设计上全部采用可循环材料，低耗能LED灯让老虎栩栩如生。

The Giant Digital Origami Tigers fuse ancient lantern making methods with cutting edge design and fabrication technology, bringing tradition and innovation, east and west together. The crouching tigers are the size of a truck at 2.5-meter high and 7-meter long, yet weigh only 200kgs. Made of fully recyclable materials, aluminium and barrisol, the big cats are brought to life with low energy LED lighting.

Formative Arts

145

绿色中空体
Green Void

项目档案	Project Facts
设计：Morfae	Design: Morfae
项目地点：悉尼	Location: Sydney

这个项目是一个临时性的雕塑装置。整体成三维的轻量合成结构，表面张力强大，不仅可以延伸到中庭的最顶端，而且连接着墙壁、天花和地板，在空中成漂浮状。绿色的表面上面依稀可见铝制的"茎"。

The Green Void is a temporary sculptural installation. A three dimensional lightweight synthetic structure, based on surface tension, is stretched to fill the atrium's void, connecting the walls, floor and ceiling. Floating in the void, the installation's green fabric is mechanically attached to suspended aluminium track profiles.

建筑造型和建筑结构

  贝克大街 55 号
55 Baker Street

项目档案	Project Facts
设计：Make Architects	Design：Make Architects
项目地点：英国	Location：United Kingdom
完成时间：2010	Year: 2010

本案的设计目的就是将空间和其价值最大化，为建筑、周围环境以及结构赋予更多的品牌效应。楼层高度为 3.20 米，天花用定做的冷硬横梁加以装饰，高度为 2.75 米。楼面板的外面有一个突口，增添了建筑的立体感，而且增加了很多室内空间。这个突口立面用了三块玻璃面罩，增添了建筑的动态美，而新的空间则淡化了公共与私人空间的界限。

The brief was to maximize the space and its value and provide the additional advantages of a brand for the architecture building and the considerable environmental and social benefits of recycling the structure. Utilizing bespoke active chilled beams, a floor to ceiling height of 2.75m was achieved within the 3.20m structural floor height. Locating risers outside the floor plate maximized heights while allowing easy upgrading or even radical changing of systems in the future as technology or occupancy evolves. The transformation is expressed by the three glass masks that span the existing blocks to create a unified dynamic facade. These new spaces deliberately blur the distinction between public and private.

Formative Arts

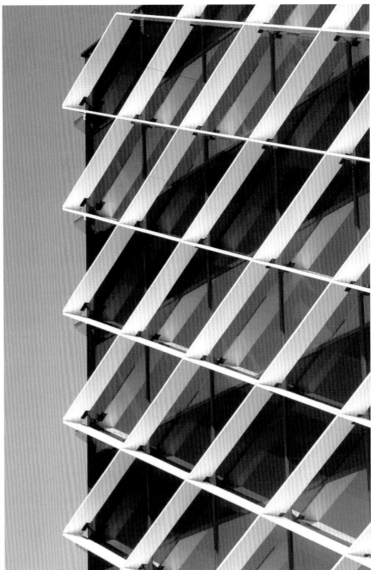

Mikrocop 数据存储办公楼
Mikrocop Data Storage Office Building

项目档案	Project Facts
设计：Groleger arhitekti	Design：Groleger arhitekti
项目地点：斯洛文尼亚	Location：Slovenia
完成时间：2011	Year：2011

这座有5 340平方米的办公楼位于斯洛文尼亚的卢布尔雅那商务区，靠近原先的工业设施。由底层和三层楼层以及平台组成的建筑用作办公室和数据保存。建筑的平面布置采用了简单的矩形，但内部分区十分灵活，所有的办公室都是透明和充满光照的，用简单的玻璃立面就可以实现这一点。两种不同的颜色展现水平的立面层，使得大块玻璃表层不再单调乏味。东南向的立面覆盖了菱形的网格，像是一张巨大的渔网，保护内部不受光照和热量的侵袭，从不同角度看，网格的空隙形成变化的视觉效果。

The floorplan of the office building is a simple rectangle with a flexible inner division, consisting of ground floor, 3 upper floors and a terrace floor reserved for offices and basement for archiving data. Providing enough transparency and light for all offices was the main goal for the architects, which led to a solution of a simple glass facade. Repeating the horizontal division of the neighbouring buildings, the facade shows horizontal layers in two different colors avoiding the monotony of the large glass surface. The east and south facades are covered with a diamond grid like a giant fishing net, which protects the interior against heat and glare. Seen from different angles, the consistent spacing of the lattice gives rise to changing visual effects.

建筑造型和建筑结构

Wykagyl 购物中心
Wykagyl Shopping Center

项目档案	Project Facts
设计：Cooper Joseph Studio	Design：Cooper Joseph Studio
项目地点：美国，纽约	Location：New York, USA
完成时间：2010	Year：2010

本案是一个翻新项目，新建筑具有全新的建筑特点，连贯优雅的立面，让整个建筑更加醒目。原来的建筑面积是 25 000 平方英尺，改建的建筑在此基础上增加了 13 500 平方英尺，包括照明设备的更新、停车场附近的绿带和景观带。通过利用有限的调色板、波浪形的立面以及铝制条带，设计师们创造了一个综合型的建筑。

弯曲的檐口形状设计增加了结构的稳定性，0.125 英寸厚的铝制条带折转成支架，这样就节省了一部分建筑材料。建筑主要采用钢制结构，表面用铝板覆盖。正立面的设计更为丰富，弯曲的铝制条板优雅地环绕在玻璃窗户上面，既能遮光又起到保护作用。

Formative Arts

The project proposes an entirely new architectural character, featuring coherent graphics and enhancing the public visibility of the center from the village main street. Renovations and additions added 13,500 square feet to the 25,000 square feet building, including lighting upgrades, new striping and landscaping of the car park and other site improvements. The designers created a unified building using limited palette materials, careful massing and a mix of new variations relating to the innovative, undulating facade treatment of aluminum banding.

The curved cornice shape gives it integral structural stability which made of 0.125 inch-thick bands of anodized aluminum bent over brackets, allowing the architects to use very little building materials. The new steel frame addition is skinned in corrugated aluminum panels. The front facade is richer in concept by the use of perforated, corrugated aluminum over the upper storey windows as a sunscreen and to protect office workers from the street at night.

建筑造型和建筑结构

 More Formative Arts

Formative Arts

建筑造型和建筑结构

More Formative Arts

Formative Arts

建筑造型和建筑结构

More Formative Arts

Formative Arts

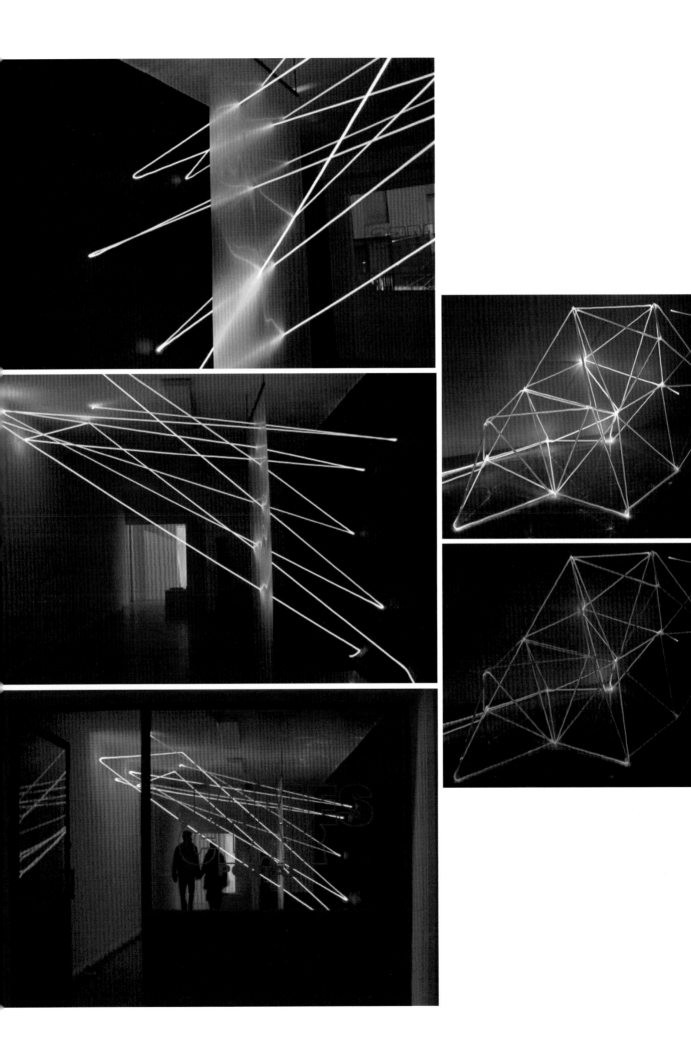

157

建筑造型和建筑结构

More Formative Arts

Formative Arts

建筑造型和建筑结构

More Formative Arts

Formative Arts

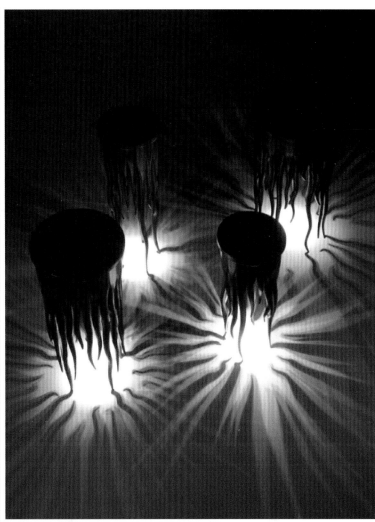

建筑造型和建筑结构

More Formative Arts

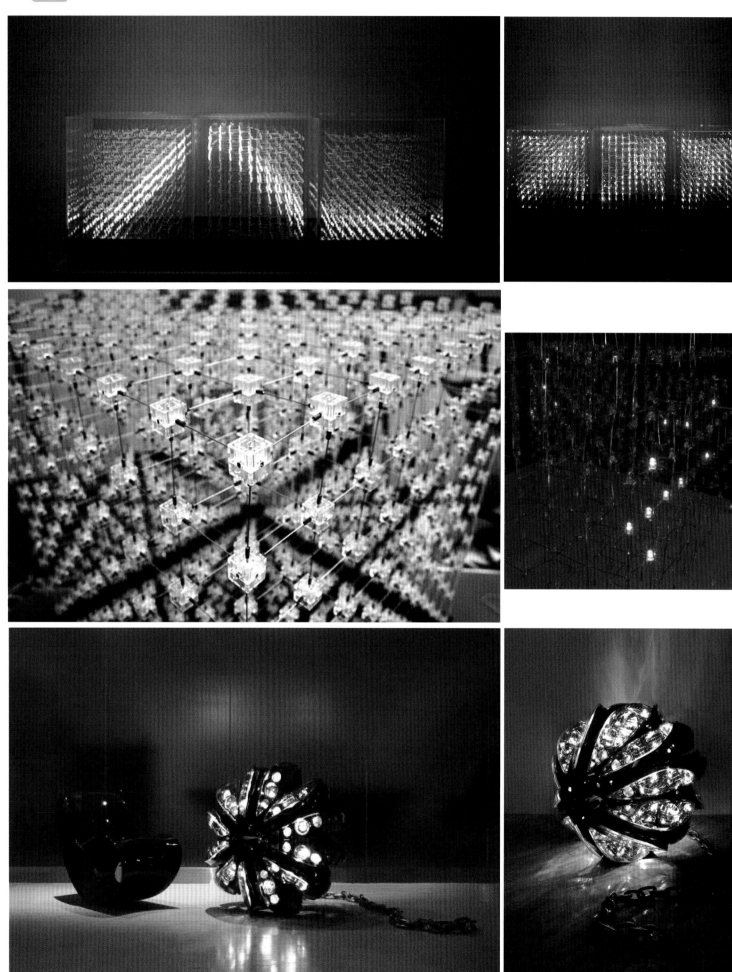

Formative Arts

More Formative Arts

Formative Arts

建筑造型和建筑结构

More Formative Arts

建筑造型和建筑结构

More Formative Arts

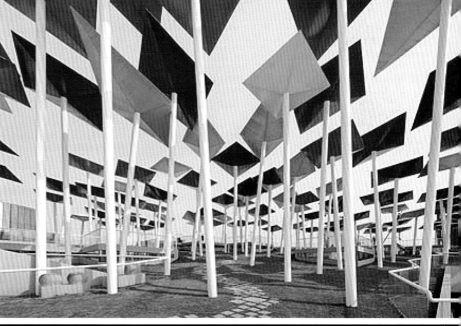

观景台 Observation Platform

黄色树餐厅
Yellow Treehouse Restaurant

项目档案

设计：Peter Eising & Lucy Gauntlett
灯光设计：ECC Lighting & Furniture- Renee Kelly
项目地点：新西兰，奥克兰

Project Facts

Design： Peter Eising & Lucy Gauntlett
Lighting： ECC Lighting & Furniture-Renee Kelly
Location： Auckland, New Zealand

这个不寻常的树屋餐厅的设计理念，是新西兰黄页公司的一个创意。建筑元素具体表达了一个椭圆形"有机体"包裹着树干，在顶部和底部，结构与结构相连，靠后的地面标高提起使圆形平面分为两部分。进入树屋前需经过一个好玩的树顶走道，访客从裂开两半的建筑裂缝中进入内部，同时体验封闭的空间与缝隙外开阔的风景这两种不同的感受。入口对面还设有一个朱丽叶露台，可以俯瞰山谷的景色。

The idea was to source all products and services through Yellow Pages listings (the book, online and mobile). The concept is driven by the "enchanted" site which is raised above an open meadow and meandering stream on the edge of the woods. The tree-house concept is reminiscent of childhood dreams and playtime, fairy stories of enchantment and imagination. It's inspired through many forms found in nature. It is also seen as a lantern, a beacon at night that simply glows yet during the day. It might be a semi camouflaged growth, or a tree fort that provides an outlook and that offers refuge. The plan form also has loose similarities to a sea shell with the open ends spiralling to the centre.

蒂罗尔山顶观景台
Top of Tyrol

项目档案

设计：Astearchitecture 建筑工作室
位置：奥地利，蒂罗尔州，斯杜拜冰川

Project Facts

Architects：Astearchitecture
Location：Stubai Glacier in Tyrol, Austria

提供各种登山和远足机会的斯杜拜冰川，距离因斯布鲁克一个小时的车程。这个新的直立的观光平台建立的主要目的是为了复苏季节性观光旅游事业。
Isidor 海拔为 3 180 米，要想到达这个观景台，首先要通过索道，然后再走过 70 米长的自然景观带。通过现场设置，并夸大现有的地形景观生成结构，换言之，这是人造景观的创造。由于含铁量高的岩石有一个红色色调，它们很显然表现出了锯齿和质感，这给了观景台独特的表现性质。游客在登高的同时能够享受无穷宽广的视觉冲击，领略静寂和休闲的宁静。

One hour's drive from Innsbruck, the Stubai Glacier offers a large variety of mountain climbs and hikes. The aim of the newly erected platform is therefore mainly the revival of seasonal and summer tourism.
The mountain station Schaufeljoch at 3,180 metres above sea level is reached via the mountain station. The path to the mountain peak platform starts from the funicular. One climbs up a number of steps to the ridge leading to the Great Isidor. After another 70 meters walk through natural landscape one arrives at the platform.
Only by creating access has a panorama view become feasible enabling the onlooker to grasp the dimensions of the landscape. The platform invites the visitor to take a rest and enjoy the peace and beauty of the mountains.

Observation Platform

穆尔河瞭望塔
Observation Tower on the River Mur

项目档案

设计：terrain: loenhart & mayr
项目地点：奥地利，施泰尔马克州

Project Facts

Design：terrain：loenhart & mayr
Location：Styria, Austria

设计呈现出的雕塑感结构是来自慕尼黑建筑事务所 terrain: loenhart & mayr。这个位于穆尔河的瞭望塔将游客带入周围森林的生态环境之中，并使游客感受附近水库的水量变化。
瞭望塔整体结构成双螺旋状，从森林中直立而起。游客们可以一边攀爬一边欣赏 360 度无限的风景。环绕形的阶梯共有 168 个，高度为 27 米。游客下去的时候是通过另外一系列的阶梯。这样，上来和下去的游客就可以避免冲突，可以欣赏到不同的风景。双螺旋状阶梯将空间和时间紧密联系在一起，就像一个螺丝。本项目的设计重点就是将空间和游客上去和下来的亲身体验相协调。

The design is the sculptural structure set amidst the landscape of the European habitat system "Green Belt".
With the look-out on the banks of the River Mur, the observation tower offers access to the ecology of the surrounding floodplain forest and lets visitors experience the river catchment, which changes according to the intensity of the water's flow. The access and construction principle of the Mur Tower is based on the idea of a double helix that is perceived as a continuous path rising up through the trees. The visitors' climb to the top is a scenic experience. The circular path, ascending to the top like a screw, passes through the different levels of the forest. The ecological storeys of the floodplain forest enables visitors to experience the eco system and the microclimate of the forest. Eventually, after 168 steps, at a height of 27 m, the observation platform is reached. A second flight of stairs leads down from the platform so that ascending and descending visitors are actually moving through the defined space on different flights of stairs. The double spiral staircase connects space and time like a screw. The connection between space and the experience of climbing up and down is the basic idea behind the spiral-shaped paths of the Mur Tower.

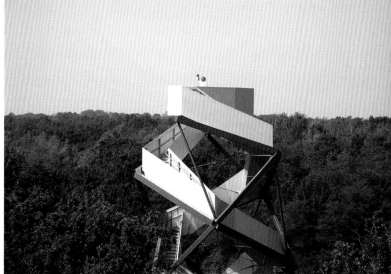

Observation Platform

观景台

Woolwich 观景台
Woolwich Lookout

项目档案

设计：ASPECT 工作室，Ducros Design, Lighting Art + Science
项目地点：澳大利亚，悉尼

Project Facts

Design：ASPECT Studio, Ducros Design, Lighting Art + Science
Location：Sydney, Australia

这个观景台是 ASPECT 工作室为悉尼港信任联盟设计的。设计理念意在构建一个可以诠释和展现 Woolwich 砂岩的公共空间，以及提供一个可以欣赏港口美景的地方。项目包括一个悬崖上的悬臂，让人们享受到无与伦比的观景体验。项目中所用的材料都是标准材料，比较适合临海环境。其中包括：钢梁、硬木甲板、混凝土、云母状的氧化铁保护涂层和不锈钢配件。

除了观景台，通往临近的"Goat Paddock"公共空间的入口和标识也是在项目设计中。入口处的大门由厚钢板构成，与临海环境相得益彰。绿化上采用当地的草，水仙和矮灌木。之所以有这样的选择，是因为它们半野生的特点以及当地的地理环境：浅淡，贫瘠的土地和带盐的风。

Observation Platform

ASPECT Studio designed the Woolwich Lookout for the Sydney Harbour Federation Trust (SHFT). The concept was to create a public place that interpreted and revealed the form of the Woolwich sandstone boat turning basin below, whilst creating a public platform for embracing distant harbour landscape views. The design of the lookout evolved to include a cantilever over the cliff edge to emphasise the sublime lookout experience. The materials palette for the lookout utilised standard materials and finishes used by the SHFT to suit the industrial maritime context. These included universal steel beam sections, heavy guage hardwood decking, mass concrete, micaceous iron oxide protective coatings and stainless steel fittings.

In addition to the lookout, new entry gates and signage to the nearby "Goat Paddock" open space were designed. These plate steel and picket gates were designed to have a significant presence in keeping with the maritime context. The planting design consisted of broad seeps of indigenous grasses, lillies and low shrubs. This planting selection was required to not impair views, to be semi wild in its character and be resilient to shallow, poor soil and salty winds.

邦代海滩勃朗特海岸步行道
Bondi to Bronte Coast Walk Extension

项目档案

设计：Ducros Design, Jeffery Katauskas, Biodesign
项目地点：澳大利亚，悉尼，澳洲维沃里公墓
完成时间：2010

Project Facts

Design：Ducros Design, Jeffery Katauskas, Biodesign
Location：Waverley Cemetery, Sydney, Australia
Year：2010

这个加高木板路创造了一个独一无二的公共景观。项目沿着东海岸悬崖边，总长515米，是South Head到Maroubra长达9 000米海岸的一部分，并与邻近的公墓相通。这条海岸线每年的游客可以达到70多万。游客们在这里不仅可以欣赏到无敌的海景，也可以体验到这里著名的人文与环境。
这个项目解决了复杂的土工技术，结构和地理环境问题，并且保护了悬崖上的湿地野生物群落。木板路的材料主要采用木材和格子玻璃纤维，使得光和水依然能供给植物。五个观景台的设施都是专门定做的，在这里人们可以驻足，休息，欣赏海岸线的美好风景。

This elevated boardwalk creates a unique public experience along the east facing cliff tops, whilst retaining unobstructed access outside the edge of the adjacent heritage cemetery. This 515m long walkway is part of the nationally significant 9km coastal walk from South Head to Maroubra that attracts more than 700,000 visitors annually. The walk is a ribbon of movement and a place to stop and embrace the heightened experience of this unique cultural, environmental and heritage landscape.
The project resolves complex geotechnical, structural and heritage conditions and retains the significant cliff-top heath community on the hanging swamps along the exposed sandstone platforms. As a direct result, the materials of the boardwalk change from timber to gridded fiberglass, allowing light and water through to remnant vegetation. Five lookout points with bespoke furniture create opportunities to pause, rest and enjoy the spectacular views along the sandstone coast.

Observation Platform

观景台

博特尼港观景台
Port Botany Lookout

项目档案

设计：Choi Ropiha Fighera
项目地点：澳大利亚，新南威尔士，博特尼港
面积：85 平方米

Project Facts

Design：Choi Ropiha Fighera
Location：Botany Bay, NSW, Australia
Site Area：85 m²

这个观景台位于博特尼海滩公园西端。项目在设计上充分利用原有的防波堤，并将其融入设计中。防波堤的边缘设计成波浪形状，一方面使得整个形状不至于太僵硬，另外一方面更好地呼应了海滩。表面经过铺装和装上甲板，方便游人的进出。这样一来，防波堤就成为了一个观景台。在这里，人们可以从不同角度欣赏美景，同时也为行人和骑单车的人提供了一个驻足点。人行道和观景台由一个适当坡度的斜坡相连。观景台的护栏是倾斜的，能够提供 270 度视角的海景。除了护栏，还有座位，以供人们休息。这样的设计想法来源于"避风向阳处"的边缘设计，这种设计在悉尼的很多海滩上都有用过。在建造材料上，选用耐用自然的，例如，回收的实木和柯尔顿耐腐蚀钢铁和混凝土，因此表面是光滑的，而且维护简单。

The Millstream Lookout is located at the western end of the foreshore beach park at Port Botany. The concept makes use of an engineering rock groyne by manipulating the structure to become part of the landscape composition. The edges are contoured to soften the form and to provide a more integrated edge to the beach whilst the surface is paved and decked to facilitate public access. This gesture transforms this utilitarian structure into a landscape element that provides a lookout, provides opportunity to view the water from a different perspective, and provides a termination to the foreshore pedestrian and cycle path. The lookout experience begins at the end of the pathway with a ramp that climbs at an accessible gradient to the lookout area. The lookout edges lean toward a view corridor out to the ocean, but at the same time provide a 270 degree vantage point of Botany Bay. Interpretive signage is integrated into the balustrade structure to coincide with the view directly in front of the viewer. Seating is also provided to allow walkers and cyclists to stop and rest. The design is inspired by the natural "sun trap" edges that exist along many of Sydneys beaches, landscape escarpments and the bow form of ships that traverse the waters of Port Botany. The construction features durable natural finishes such as recycled hardwoods, "Corten" steel and textured concrete so that the structure will age gracefully whilst also requiring a lower level of maintenance.

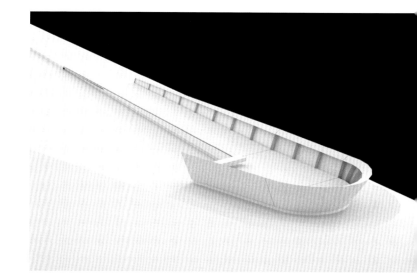

Observation Platform

挪威 Ornesvingen 观景台
Ornesvingen

项目档案

设计：Smedsvig Landskap AS, May Elin Eikaas Bjer
项目地点：挪威，盖伦格峡湾

Project Facts

Landscape Architecture：Smedsvig Landskap AS, May Elin Eikaas Bjer
Location：Geiranger, Norway

盖伦格峡湾，被联合国教科文组织公认为世界文化遗址，也是挪威最著名的旅游胜地。陡峭的峡谷，弯曲的公路，这个项目的观景台位于其中一处拐弯的地方，是游客欣赏盖伦格峡湾最壮观的观景台。

整个项目包括三块重叠的混凝土厚板，下空高度为600米。观景台从悬崖边缘延伸出去，游客们可以悬空欣赏风景。旁边的河水经过特别设计，流过光滑的表面，在厚板边缘形成一道天然的瀑布。用玻璃做成的信息板为游客们提供了具体的位置信息。

Ornesvingen is the most spectacular viewpoint along a zig-zag road along the steep valley sides of Geiranger-fjord. Located at one of its many bends, the viewpoint gives tourists breathtaking views over the recently acknowledged UNESCO world heritage site, making it one of Norway's major tourist attractions.

The project consists of three overlapping white concrete slabs overhanging the edge of a 600-meter vertical drop, enabling the observer to step out into the airspace. The river on site is guided over a glazed front, forming a waterfall on the very edge of the viewpoint. Information boards in glass explain specific locations in the scenery.

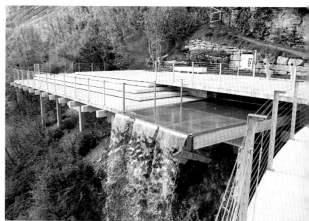

Observation Platform

185

观景台

Tungeneset
Tungeneset

项目档案

设计：Aurora Landskap
项目地点：挪威，Senja

Project Facts

Landscape Architecture：Aurora Landskap
Location：Senja, Norway

这个项目位于一个用作停车场的空地上，和公路直接相连。80 米长的结构直接通向烧烤区域。结构的栏杆不仅成为这个区域的标志，而且限制了车辆的停靠，减少了这里的污染。结构的前 35 米是一个混凝土构成的沿着公路和海岸岩石的斜坡。在其尽头，有一个小型的观景台，在天气恶劣的时候，这里是最安全的地方。再往后就是一个横跨在岩石上的木桥。木桥上不规则的栏杆与周围的地形遥相呼应。木桥的尽头就是烧烤区，在这里有很多不同的座位，满足各种不同的需求。

The project starts at a carpark directly by the road. The railings by the carpark is lit up in winter to mark the area, and also to limit the use this time of year. From there a 80m long structure makes its way down towards the barbacue area, located on the costal rocks. The first 35m of the structure is a concrete ramp following the slope between the road and the coastal rock. At the end of the concrete ramp, there is a small viewpoint, the safest place to stop on days with bad weather. From there the structure continues as a wooden bridge floating over the rocks. The bridge as well as its irregulare railings is an echo of the way it moves through this type of terrain. The brigde ends in the barbacue area, where different sitting arrangements gives room for several ways of use.

Observation Platform

灰色的混凝土和岩石是相同的颜色,看起来像是从岩石上抠出来的一样。

The grey concrete has the same colour as the rocks, and looks like it is cut out of it.

观景台

Flydalsjuvet 峡谷卫生间
Flydalsjuvet

项目档案

设计：Smedsvig Landskap AS
项目地点：挪威，盖伦格峡湾
面积：170 平方米

Project Facts

Landscape Architecture: Smedsvig Landskap AS
Location: Geiranger, Norway
Site Area: 170 m²

Flydalsjuvet 峡谷位于盖伦格峡湾末端，公共卫生间则建在峡谷里一片依山的陡峭地面上。设计事务所用有几百年历史的木房屋上拆下的木材作厕所和信息栏的框架。设计师们从当地历史悠久的地方搜集了这些木材，雇佣传统工匠对这些木材进行翻新。他们将木制墙壁安在建筑玻璃框架结构上，这样自然光就可以射进室内空间。这三间独立的建筑建在成阶梯状的混凝土基座上，设计师用极具前瞻性的设计和建造工艺对当地传统建筑方式表示敬意。

Flydalsjuvet is a canyon at the end of the Geiranger Fjord. The resting area is located on one of the steep mountain sides leading to the Fjord, across the valley from nesvingen. Several-hundred-year-old timbered building modules are reused as framework for a new structure to facilitate toilets and information stands. The logs are collected from a local site and refurbished by tradition-carrying craftsmen. The walls are then mounted on a structural glass base, allowing light to enter under the massive wooden walls. This is an homage to an old local building tradition preserved for the future, floating on a modem glass construction. The three individual buildings are placed on a terraced concrete floor with narrow pass-ways between them leading to the ridge of the mountain.

Observation Platform

观景台

阿尔塔米拉诺散步道
Paseo Altamirano

项目档案

设计：Emilio Marin，Nicolas Norero
项目地点：智利，瓦尔帕莱索

Project Facts

Design: Emilio Marin, Nicolas Norero
Location: Valparaiso, Chile

该项目位于科尔多瓦观景台与另一个名为"绝壁"的观景台南部之间，靠近T1隧道，是进入瓦尔帕莱索市新的南门，它在瓦尔帕莱索市市郊一片被废弃多年的土地上，这里只有土路和悬崖。这个地方介于城市与大海之间，为该市提供了一个相对宽广的空间，不仅可以用来填埋垃圾，而且还构成了该市的第五扇大门。这使得该项目成为阿尔塔米拉诺散步道 (Altamirano Walk) 上一道亮丽的风景，同时还形成一种小范围的生态平衡系统。而该项目通过使用与瓦尔帕莱索这个历史城市不直接相关的设计素材，在海岸沿线创造了一种别样的风景。

设计师们用简单的预制六角形混凝土板作为整个项目的根基，同时在上面还做了许多富于变化的纹理和结构，其目的就是要将瓦尔帕莱索的新南门海岸沿线引向繁荣。

The project is situated between the Cordoba Viewpoint and a field south of another viewpoint called "El Acantilado" (The Cliff) and close to the T1 tunnel, the new south entry to the city of Valparaiso. This place between the city and the ocean presents a special opportunity to provide a relevant space, not only for its plastic proposal, but also because it complies and draws together all of the challenges set forth by the competition: integral unity of the components and a dimension that corresponds to a "fifth facade". This enables the project become an active part of the walkway and at the same time it captures a climate within it. The project sets out to re-found the place using elements that are not directly related to the historical city of Valparaiso but that can embrace different possibilities along the coast.

A simple prefabricated hexagon-shaped concrete slab becomes the constructive base for the project, with variations of texture and composition. The shades follow the same principle dictated by the hexagon, maximazing the commercial format of the material which, in this case, is steel.

In a final stance, this project aims to bring value to the coastal line of a new southern entry point for Valparaiso which has been declared a World Heritage site by UNESCO.

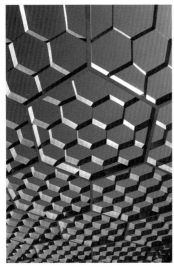

Observation Platform

观鸟台——挪威旅游线路项目
National Tourist Routes Projects

项目档案

设计：70'N 建筑事务所
项目地点：挪威

Project Facts

Design：70'N Arkitektur
Location：Vestvagoy, Lofoten Islands, Norway

塔的入口被高墙和观景台屏蔽着，因此观鸟者的影子不会打扰到小鸟们。这座塔是钢筋混凝土结构，外部用原木包裹着。在地面上有一间狭窄的玻璃房，上面一层有大面积的区域，方便游客欣赏风景。塔的稳定性是很重要的，这座塔能够抵挡住强风，还不影响游客们使用望远镜。北面还有一个单车棚，这里也是游客们暂时避风的好地方，而且能让游客们近距离接触真正的大自然。

The entrances to the towers have been screened off with high walls and the observation platforms are formed so that no silhouettes of the bird-watchers are cast in order not to disturb the birds during the breeding season. At the entrance level, there is a weather-protected room with a narrow glass observation opening. The upper level has large open areas for the best possible views. The tower is a robust steel construction with secondary wooden construction of untreated wooden fronts. The stability is important, so that the tower can withstand strong winds without affecting vibration sensitive binoculars. The bike shed is situated at Grunnfor on Austvagoy in northern Lofoten with an open northwards view towards Vesteralen and a grand southwards view towards the mountains in the south. Here the visitor can seek shelter from the wind, which can be extremely cruel, and also have a magnificent experience of the nature.

Observation Platform

水袖天桥
Long Sleeve Skywalk

项目档案	Project Facts
设计：Turenscape	Landscape Architecture：Turenscape
项目地点：徐州市，遂宁	Location：Suining County, Xuzhou City
长度：860 米	Length：860m
完成时间：2010	Year：2010

水袖天桥简单而优美，以诗情画意的方式将城市的复杂功能表现出来。天桥横跨几条河流和一条高速公路，并且连接着和谐广场和森林大厦。主桥长度为635米，建造长度为869米，总面积为2 700平方米。四座辅助的桥总长为242米，甲板宽度为2.5-9米，坡度为0.4%-12.6%。天桥的高度为4.5米。

这个项目以加强和谐广场和森林大厦的连接，保证行人的安全性为主要目的，通过分开和加高人行通道，来避开徐宁路繁忙的交通影响。在满足基本要求的同时，设计想法主要来源中国京剧中的水袖。另外，这个项目的复杂照明系统非常引人注目。晚上，灯光柔和，水袖天桥显得更加楚楚动人。水袖天桥城市景观功能和形式在艺术上统一的典型案例，将城市设施转化成一个艺术性景观。

Observation Platform

观景台

Observation Platform

The Long Sleeve Skywalk is a simple but graceful integration of complicated urban functions with poetic space. Straddling across several river systems and a express way (Xuning Road), the Long Sleeve Skywalk is located on Xuning Road, Suining County, Jiangsu Province. It connects the focal point of the county, Harmony Square, to the Forest Plaza across the road. The main bridge spans 635m, with a total construction length of 869m, and covers a total area of 2,700 m^2. Four auxiliary bridges span a total length of 242m, with decks ranging from 2.5m to 9m wide, sloping from 0.4% to 12.6%. The bridge spans a net height of 4.5m over the road pavement below.

Based on the initial premise, the bridge is designed to strengthen the spatial connections between Harmony Square and Forest Plaza to ensure safe pedestrian crossing. This is achieved by separating and elevating the pedestrian pathway over fast moving traffic on Xuning Road. Beyond meeting the basic requirements, the design was inspired by the dancing shadows of long sleeves in the Chinese operas like Peking Opera. In addition, the bridge's intricate light display system glows softly at night as it weaves fluidly between the city square, waters, and woodland.
The Long Sleeve Skywalk sets an example of the artful integration of function and form of urban landscape elements, and turns urban infrastructure into an artful landscape.

观景台

海边景观
Microcostas

项目档案	Project Facts
设计：Guallart Architects	Design：Guallart Architects
项目地点：西班牙，Vinaros 南海岸	Location：South Coast, Vinaros

Vinaros 是西班牙紧邻地中海海岸的一个城镇。南面的海岸由一系列连续的海湾和海岬构成，而海湾和海岬主要由断层的砾岩构成。随着大海的四季变化，海岸线的长度和周边城镇的面貌都会有变化。这个项目的目的就是在小地块上建造一个个分离开的房子。

这个项目扩大了这个海岸的视线范围。从环境或城市方面来说，这个海岸原本没有多少利用价值，因为不适合城市居住的发展。一眼望去，海湾缺少了自然的韵味；但是细看，你却会发现经海潮洗礼后独一无二的美。

六边形的木制平台边缘的长度是依据人的身体特点来规定的，平台由两部分构成，一部分是平坦的，一部分是向上延伸开来折叠起来的；这种平台适合于任何大小的海岸。在考虑这些平台的设计地点的时候，设计师充分考虑到海岸的进出路线以及与动态海岸线的互动性。这样一来，到这里游玩的人们就可以很快地适应新的海边风景，并充分发挥这些平台的各种用途。基于空间的社会学形态，平台的尺寸，走向和地点之间的关系值得人们深思。

Vinaros is a town on the Mediterranean coast of Spain. Its south shore is a succession of coves and promontories on a terrain composed of strata of easily fractured conglomerate rocks. The length of the coastline and the surface area of the municipality are constant changing as a result of the action of the sea, which produces continual land slippage and erosion. This zone has been developed with detached houses on small plots.

This project can be taken to exemplify the way the scale of the gaze is the key to perceiving the logic to be acted on. At the intermediate scale the place is of very little interest in urbanistic or environmental terms, given the proximity of the residential developments to the coast. The coves and points appear at first sight to be far removed from the ideal "virgin" state of what could be described as natural. On the small scale, however, it becomes clear that this sequence of coves and outcrops, micro-inlets, pools of seawater, stones eroded by the sea and rocks shaped by the tide have an exceptional beauty.

Observation Platform

The project has consisted in establishing a mechanism to measure the coast, on the basis of the creation of hexagonal timber platforms with a constant length of side based on the scale of the human body. These micros-coasts are organized to form islands of variable size, located where there is rock in dose proximity to the sea. The platforms are composed of just two different pieces, one flat, the other with a microtopography, which serve to generate surfaces that can be perfectly flat or partially or fully folded. Their positioning on the coast is determined by criteria of access to the sea and interaction with the dynamic line of the original coast. Following their installation, people are quick to be appropriate the new micro-coasts and utilize them in a variety of ways. The relationship between the size, orientation and location of the platform and the number and social profile of the people using them is an interesting phenomenon in terms of the socialization of the space.

观景台

Gudbrandsjuvet 观景台
Gudbrandsjuvet

项目档案

设计：Jensen, Skodvin
面积：350 平方米

Project Facts

Landscape Architecture：Jensen, Skodvin
Site Area：350 m²

观景台主平台从悬崖向外伸出呈悬臂式，是由 25 毫米的激光切割薄钢板搭建而成的。几何结构的栏杆在适应不同地区安全性要求的同时还保证了景观整体的连续性。栏杆呈向内侧弯曲的曲线，使游客能安全、尽情地探身欣赏美景。人行桥梁根据不同区域的不同特色由不同的材料搭成。停车场旁边的平台由混凝土预制构件搭成，形似自行车链条，与周围环境相适应。悬臂式预制构件无论是从经济角度还是实用的角度都是最适应此处环境的。相关的几何学原理同样应用于服务中心。

The main platform is constructed by 25mm laser cut steel sheets, cantilevered like a bridge around the cliff, hung in each end. The railing has a geometry that allows it to be continuous even with very different security requirements from place to place. The large inward curve allows the tourists to securely lean out over the deadly waters. The bridges are made from different materials according to what is most appropriate at each site. The platform at the parking side is made from prefabricated elements of concrete, like a bicycle chain, an element that is connected in the comers but rotated in the angle that will fit the site. This is appropriate at this site because cantilevering prefabricated elements have obvious advantages economically and practically. A related geometric concept is used for the service center.

Observation Platform

观景台

Trollstigplataet 观景台
Trollstigplataet

项目档案

设计：Reiulf Ramstad 建筑事务所
项目地点：挪威
面积：150 000 平方米

Project Facts

Design: Reiulf Ramstad Architects
Location：Norway
Site Area：150,000 m² (landscape)

这个项目是"国家旅游路线"建设的一部分，将会有助于游客更好地体验特罗斯蒂高原，让他们欣赏到无与伦比的自然风景。项目的设计风格以及所选择的材料都适合当地的特色，功能设施上也能促进游客的亲身体验。建筑与周围景观之间有清晰的精确的过渡。积雪融化成水，作为动态的元素，从高山上流下。另外还有石头这一静态元素，形成鲜明的对比。这个项目的空间性是独一无二的。

Project is a part of a "National Tourist Routes" project. The project will enhance the experience of the Trollstigen plateaus location and nature. Thoughtfulness regarding features and materials will underscore the site's temper and character, and well-adapted, functional facilities will augment the visitors' experience. The architecture is to be characterized by clear and precise transitions between planned zones and the natural landscape. Through the notion of water as a dynamic dement from snow, to running and then falling water, and rock as a static element, the project creates a series of prepositional relations that describe and magnify the unique spatiality of the site.

Observation Platform

观景台

Observation Platform

观景台

洛格罗尼奥鸟类观测台
Crnithological Observatory

项目档案

设计：Manuel Fonseca Gallego
项目地点：西班牙，洛格罗尼奥

Project Facts

Design：Manuel Fonseca Gallego
Location：Logrono, Spain

在成熟的城市环境中建造一个鸟类观测台，作为一个独特而突出的景观标志，这想法源自于发起人。他曾提出不少风景非常优美的地方作为观测地点的建议。经过详细的帧摄影分析，最终定在一个我们认为最为合适的观测据点，既可对大量候鸟进行细致观察又不会破坏其自然栖息地。具体的位置是在两棵大树之间，这样可保证建筑体量充分与河畔植被融为一体。显而易见，木材是最合适的建筑材料，同时它也构成了最有趣的建造部分。

项目的主要概念是将建筑视为一个大的箱形梁，由四个翼面规划出体量，从而形成直纹曲面造型。这些翼面是由木板不规则排列而成的，木板覆盖了部分的体量表面，体型像一个"陷阱"。走进这个鸟巢似的世界，便可感受到光影交错的超现实氛围。

The idea of an ornithological observatory, a unique and eminently landscaping object located in a well established urban surrounding belongs to the promoter, who proposed several locations within an area of outstanding beauty. After a detailed frame photographic analysis, we chose the most suitable location we considered among other reasons, because it's a strategic point for the detailed observation of a large number of migratory birds without altering their natural habitat. The rest consisted in locating it between two magnificent tree specimens to ensure that the volume was fully integrated with the riversides vegetation as an element itself. The choice for the material seemed clear. Wood was the most interesting construction component.

Observation Platform

The main concept is to consider the item as a big box beam bounded by four planes, the lower horizontal one and the rest formed by ruled surfaces. The constructive composition of these planes is set with wooden planks arranged irregularly, covering the surface partially, as a "sell". We dip ourselves into the dream world of nests, recreating a surreal atmosphere with lights and indoor shadows.

标识
Signage

美国匹兹堡儿童博物馆
Children's Museum of Pittsburgh

项目档案	Project Facts
设计：Koning Eizenberg	Design：Koning Eizenberg
项目地点：美国匹兹堡	Location：Pittsburgh, USA

美国匹兹堡儿童博物馆以全新的大门造型欢迎各方来宾。超大型的钢铁结构架起的门廊，配上一架秋千，以一种全新的形式演绎了传统的欢迎符号。走过门廊，踏入大门，仿佛进入了一个生动而充满各种可能的世界。以"摆弄真实的器具"的理念为基础的互动式展览一字排开并一路从旧邮局通道延伸到博物馆的左侧；而在博物馆的右侧，门廊则变身为一个旧式的布尔细工入口大堂。这个空间安装了大量大块的玻璃，方便照明。同时在视觉上，将三栋建筑联系起来，让儿童与户外建立空间联系，这也是教育场馆设计中非常重要的儿童发展课题。

The Museum welcomes the public through its new front door. An oversized steel-framed verandah, complete with a porch swing, becomes the old symbol of welcome in a new form. Passing under the verandah and through the entry, one immediately steps into a world of activity and possibility. Interactive exhibits based on the philosophy of "Play with Real Stuff" are straight ahead and through the old Post Office tunnel to the left. To the right, the verandah is converted into the old entry lobby of the Buhl, a beautiful volume now enhanced by a dramatic new window for much-needed light, visually linking all three buildings and allowing children to establish their spatial connection to the outside world, which is also a critical child development issue in institutional design.

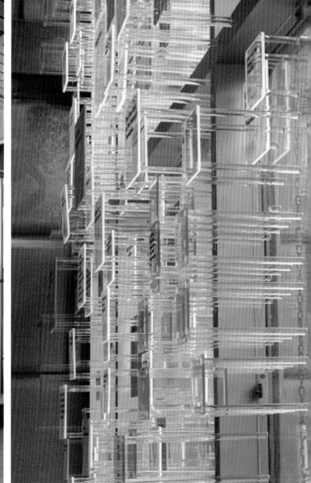

Signage

标识

 Halle F 音乐厅指示牌
Halle F Signage

项目档案　　　　　　　　Project Facts

设计：Justus Oehler　　　Design：Justus Oehler
项目地点：奥地利，维也纳　Location：Vienna, Austria

Justus Oehler 和他的设计团队为 Halle F 音乐厅设计了标识牌和指示牌。该音乐厅位于维也纳 Wiener Stadthalle 场馆内，可容纳两千名听众。

Justus Oehler and his team have designed the identity and wayfinding signage for Halle F, a new 2,000-seat concert hall at the Wiener Stadthalle in Vienna.

标识

马里兰艺术学院
Maryland Institute College of Art

项目档案　　　　　　Project Facts

设计：Flux Studio　　Design：Flux Studio
项目地点：美国，巴尔的摩　　Location：Baltimore, USA

马里兰艺术学院的标志性设计是马里兰艺术学院入口大楼的灯光。它不仅强调了建筑的动态外形，也表现了正在发生的各种活动。从外面望去，整齐划一的窗户覆盖了整个建筑体，由内而外散发出玻璃的光泽，因室内的活动而形成各种不同的图案。灯饰树排列在皇家山和北大街两旁，强调了建筑作为校园夜间出入口的地位。可调节的剧场式灯具为主要的外部空间提供了基本的光源，并为在此举办的各种活动提供多种灯光效果选择。这些灯光还参考了大楼一层的实验剧场的灯光效果设计。

Signage design for Maryland Institute College of Art is lighting for the gateway building, which is designed to accent the dynamic architectural forms and express the exciting goings on the site. When viewed from around the city, the modular window system covers most of the building exterior glows from within, and creates different patterns based on resident use. Illuminated trees line Mount Royal and North avenues reinforcing the intent that this project serves as a gateway to the campus at night. Adjustable theatrical light fixtures are the primary source of light in the major exterior spaces. In addition to offering flexibility in creating dramatic lighting effects for the variety of events that will take place in these locations, the objects themselves refer the black box theater located on the first floor. The lighting design also features an urban scale illuminated LED sign on the upper floors of the north side of the building.

Signage

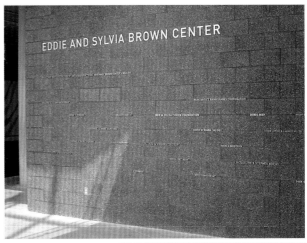

标识

新教学大楼
New Academic Building

项目档案　　　　　　　　　　　　Project Facts

设计：Abbott Miller, Jeremy Hoffman　　　Design：Abbott Miller, Jeremy Hoffman
项目地点：美国，纽约　　　　　　　Location：New York, USA

新教学大楼位于库珀广场41号，隔着第三大道，正对着柯柏联盟学院旧有的于1859年建成的基金会大楼。和Mayne的建筑设计一样，Miller为新大楼所做的墙面设计也和旧有建筑之间，建立了某种对话。至于标识文字的印刷，Miller选择了Foundry Gridnik字体，和基金大楼外墙上的字体相类似。原有的标识牌有着棱角分明的硬朗形象，代表了艺术、建筑和工程。标识的印刷通过各种方式呈现出来，立体标识牌表面的文字呈现出被挤压、切割、延伸或拖拉的效果。Miller和Morphosis密切合作，将标识牌融入到建筑当中。建筑顶盖采用了光学冲压而成型的字母，在某个严格的高度上看，字母是"正确"的，可一旦在空间上往后弯曲时，字母的外形就产生了扭曲。字母下半部分的镂空与穿孔不锈钢制成的建筑表皮的通透感相呼应。Miller设计的捐款人姓名标识同样以独特的方式与建筑连接起来。新建筑的大堂是一个自然采光的中庭，贯通建筑核心部位，有9层楼高，主要结构是楼梯。一个标明了主要捐款人名字的大型结构让楼梯的底部活泼起来。该标识由80片"刀片"组成，位于楼梯的底部，随着楼梯的走势往下排列。捐款人的名字被印刷在每片"刀片"的正面、底面和反面。

同样的印刷方式也运用在教室和办公室门外所安装的捐款人标识牌上。房门有10英尺高，而门外的每块捐款人标识牌就好像门外的一条竖立的不锈钢扶角铁条。铁条上的印刷文字仿佛像是被人从一侧拖拉到另一侧，所以，标识条的一侧看起来像是覆盖了黑色条纹，而另一侧则可以看到字母。

Signage

标识

Signage

The new academic building, located at 41 Cooper Square, sits directly across Third Avenue from the Cooper Union's original 1859 building, called the Foundation building. Like Mayne's architectural design, Miller's graphics for the new building establish a dialogue with the older structure. For the signage typography, Miller chose the font Foundry Gridnik, which resembles the lettering on the facade of the Foundation building. The original signage has a strong, angular look that suggests art, architecture and engineering. The signage typography has been physicalized in different ways, engaging multiple surfaces of the three dimensional signs, appearing extruded across comers, or cut, extended and dragged through the material. Miller worked closely with Morphosis to integrate the signage into the building. The building canopy features optically extruded lettering that appears correct when seen in strict elevation, but distorts as the profile of the letter is dragged backwards in space. The cutouts in the lower half of the letterforms echo the transparency of the building's surface "skin" of perforated stainless steel.

Millers program of donor signage is also uniquely integrated with the architecture. The lobby of the new building is a soaring sky-lit atrium that rises up nine stories through the building's core and is dominated by staircases. A dramatic installation recognizing major donors animates the underside of a descending stairwell, the signage comprised of over 80 "blades" that cascade down the underside of the stairs, echoing the stairs, downward motion. The typography is engraved on the front, bottom and reverse surface of each blade.

A similar typographic approach was used for a series of donor signs outside classrooms and offices. The rooms feature doors that are 10-feet high. The donor sign at each room's entrance appears as a vertical stainless-steel bar that acts as a "comer guard" for the doors. The typography appears as though it has been dragged through the bar, so edges of the letterforms are visible on edges of the bar. When seen from the sides, these bars appear covered in patterns of black banding, another way to see letterforms.

标识

芝加哥艺术学院
The Art Institute of Chicago

项目档案

设计：Abbott Miller, Jeremy Hoffman
项目地点：美国，芝加哥

Project Facts

Design：Abbott Miller, Jeremy Hoffman
Location：Chicago, USA

新标识的灵感来源于位于著名的密歇根大道上的 1893 艺术大楼外墙上的题字。遵循古典和新古典风格题字的作法，外墙上的标识选用了罗马字体 V 来代替 Institute 这个单词中的字母 U。新标识代表了现代钢琴式设计的建筑一翼与古典主义的密歇根大街建筑之间的平衡点。从旧立面看过去，那标识有着奠基石一般的庄重感；当从现代建筑翼那面看过去，则有着与琴键设计结构相呼应的轻盈和精确质感，标识与建筑充分结合。此外，还运用了铝材、莱姆石和玻璃，使硬质坚固元素和线性半透明材质相互融合。Miller 和他的团队同时还为整个博物馆开发了一套完整的道路指引、指南和捐款人标识系统，还在现代翼楼中部的格里芬中庭内设置了捐款人姓名墙。

The new identity was inspired by the inscription of the name on the famous Michigan Avenue facade of its 1893 Beaux Arts building. Following the practice of classical and neo-classical inscriptions, the lettering on the facade uses the Roman V in place of the U in the word Institute. The new mark strikes a point of balance between the modernity of the Piano-designed wing and the classicism of the Michigan Avenue building. When seen in relation to the historic facade, the mark has the gravitas of an engraved cornerstone, but when seen in relation to the Modem Wing, it echoes the precision and lightness of Piano's structure. The signage has been fully integrated with the architecture of Piano's addition, using aluminum, limestone and glass to provide an interplay between solid and engraved elements, as well as linear and transparent elements. Miller and his team also developed a complete program of wayfinding, directories and donor signage for the entire museum, as well as several donor walls in the Griffin Court, the soaring central space of the Modem Wing.

Signage

标识

第五大道 623 号
623 Fifth Avenue

项目档案

设计：Abbott Miller
项目地点：美国，曼哈顿

Project Facts

Design：Abbott Miller
Location：Manhattan, USA

图为曼哈顿中心区一幢 36 层莱姆石外墙办公楼的标识牌设计。

Identity and entry signage for a 36-story limestone-clad office building in Midtown Manhattan.

标识

多伦多皮尔逊国际机场
Toronto Pearson Airport

项目档案

设计：Michael Gericke, Wayne McCutcheon
项目地点：加拿大，多伦多

Project Facts

Design：Michael Gericke, Wayne McCutcheon
Location：Toronto, Canada

Pentagram 设计公司为多伦多皮尔逊国际机场新建的一号航站楼设计了一套方便查找的道路指示系统。新建的一号航站楼由 Skidmore, Owings & Merrill 与 Moshe Safdie and Associates 公司合作设计，是加拿大历史上规模最大的公共建筑。

项目内容包括两方面的重要因素，它们分别是机场内功能区的标识牌（比如售票柜台、登机口、行李传输带等）和指示方向的吊牌。出境大厅中建有高大的标识塔，标示出办理登机手续柜台、方向、登机门号码和其他指示信息（如显示飞机升降信息的LCD 屏幕）。标识塔的形状如同空中交通管制塔和加拿大国家电视塔。而高悬的指示吊牌则有着艺术雕塑般的造型，包含了英、法指示语两个部分。指示吊牌的弧形与航站楼的穹顶以及螺旋桨的流水线造型相呼应。大量研究，包括有视力障碍人士参与的测试证明，图形在宽敞的航站楼空间里保证最大的可视性。旅客可以方便地通过按照用途（登机口、陆上交通、海关和配套设施）喷上不同颜色的象形指示牌，穿越航站楼。所有这些标识牌都由铝制成，内部装设耐用的照明设备，将设备维护工作量降到最低。

Pentagram developed a comprehensive wayfinding program for the Terminal One at Lester B, Pearson International Airport in Toronto, designed by Skidmore, Owings & Merrill in collaboration with Moshe Safdie and Associates, which is the largest public building project in the history of Canada.

Two key elements of the program were components marking destinations within the airport (such as ticket counters, gates and baggage carousels) and hanging directional signs. The departures hall contains tall pylons that present check-in-counter identification, directions, gate numbers and other information (including arrivals and departures in LCD screens mounted in the base), in a shape reminiscent of air-traffic-control towers and Toronto's own CN Tower. The overhead hanging signage is a sculptural structure composed of two sections, one for English and one for French, in curving forms that echo the arc of the terminal roof and the shape of an airfoil. Extensive studies, including tests with visually impaired users, were done to ensure that the graphics would provide maximum visibility across the terminal's vast interior spaces. Travelers can easily make their way through the terminal following messages and pictograms that are color coded according to user path (gates, ground transportation, customs and amenities). All fixtures are made of aluminum and internally illuminated with long-life lighting elements for minimal maintenance.

Signage

标识

Signage

标识

Signage

高架公园
The Highline

项目档案

设计：Michael Gericke, Wayne McCutcheon
项目地点：美国，纽约，曼哈顿

Project Facts

Design: Michael Gericke, Wayne McCutcheon
Location: Manhattan, NYC, USA

该项目位于纽约市曼哈顿区，项目将一条废弃的铁路桥改造成一个社区公园。Intrex 公司为该项目生产了大量配件，包括自行车架、手扶栏杆、种植池、花托、绿化墙体、玻璃防护栏和其他大量固定装置。项目通过设计师、景观师和建筑师的通力合作，大获成功。这个充满想象力的项目始于 2000 年，由一群毫无城市规划经验却充满热情的观察家发起，希望拯救这条废弃的纽约铁路，使其避免被拆除的命运。而这一项目目前已经成为了市内的头号旅游景点，同时也为世界各地提供了有益的借鉴。

The Highline is a project in Manhattan NYC that turned an abandoned railroad tressel into a community park. Intrex manufactured numerous items for this project, including Bike Racks, Hand Rails, Planters, Receptacles, Recyclers, Living Walls, Glass Barriers, and numerous other fixtures. Working together with the designers, landscapers, and architects, this project was an incredible success. Started in the year 2000 as a fanciful project by passionate observers, with no urban planning experience, to save an abandoned New York rail line from demolition, the project has found itself as the city's number one tourist attraction and an inspiration to councils and cities around the world.

Signage

Signage

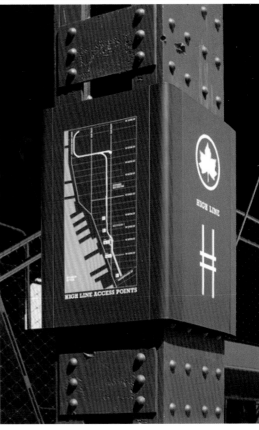

标识

Cogeco 公司总部标识
Cogeco Headquarters Signage

项目档案	Project Facts
设计：Waltritsch a+u	Design：Waltritsch a+u
项目地点：意大利，雅斯特	Location：Trieste, Italy

Cogeco 的大堂以一堵多层折叠式墙面为特征。墙面其实就是一张虚拟地图，保留了纬度平行线，线与线之间标注了来自全球各地咖啡的名字。同时通过墙面上一个巨大的咖啡因化学式，作为墙面转角的标志，也代表着公司的专业。通过这种装饰，人们瞬间就能一览"全球"，同时了解到公司的业务范围。原料鉴定实验室是公司对咖啡原料进行一系列测试并出具品质证明（烘烤、大小测定、气味和味道等），以及大部分业务交易完成的地方。咖啡样品袋放在彩色架上，包围了整个房间，整个空间再次被桌上一个折叠式的巨型地图所占据，形成详细的古德地图投影，得出各种主要咖啡种类原产地的精细地理信息，同时也为鉴定实验室增添了视觉层次。

The lobby of the Cogeco is characterized by a multilayered folded wall, an abstracted map where only the parallels of the globe have remained, which hosts a series of "exotic" coffee names coming from all over the world, and by a big chemistery formula of the caffeine, which marks the comer and gives evidence of the specific knowledge. In such a way, one is immediately transported in a quick ride through the globe, and at the same time given a clear statement of the company's know-how. The proof and taste lab is the place where the company makes a series of tests on the raw good in order to provide a certificate of quality (roasting, checking dimension and smell, tasting etc), and where most of the commercial deals are made. Coffee sample bags are exposed on a colored shelf marking the perimeter of the room. The space is again dominated by a big map folding on this table, a very detailed Goode Homolosine projection of the world, where precise geographic indication about the origin of the most important coffee types is given, adding a visual layer to the proofing experience.

Signage

美发屋临街面标识
Hairstyle Interface Signage

项目档案	Project Facts
设计：X Architekten	Design：X Architekten
项目地点：奥地利，林兹	Location：Linz, Austria

立体的建筑立面很是特别，发丝状的波浪，成为这个美发屋吸引客人的一大亮点。同时，这种动态的设计也兼具了其他的：在垂直方向上，立面看起来像是窗帘，不仅提供了一定的隐私性，而且增添了美发屋的氛围和特色；从街道上不同的角度看，发丝状的波浪都很优美，足以吸引路人。发丝状的波浪是有层压板构成的，垂直立在立面上，在恰到好处的角度上凸起。连同上端的标识，发丝状的波浪都是深深依附在立面的内部，这样就保持了和建筑轴线的关系，与周围随意的环境很好地兼容。

The entire facade now acts as an effective store sign through a three-dimensional architectural hair wave which runs along the exterior. At the same time, this dynamic design fulfils another important function: vertically, it runs like a curtain over the glazed facade, offers a degree of privacy, and thus supports the atmosphere and the character of the salon. By looking from different viewpoints in the street, the three-dimensional hair wave proves attractive to passers-by. The appearance of the flowing hair as well as the above inscription is incorporated into the facade in a way that maintains the relationship with the building's axes. In this way it harmonises extremely well with the generally uniform environment. Laminated sheets, arranged vertically with varied spacing and at a right angle to the facade, are used as the material for the hair wave.

Signage

埃斯托伊宫酒店指示牌
Palace of Estoi Signage

项目档案	Project Facts
设计：Cristina Catarino	Design：Cristina Catarino
项目地点：葡萄牙，里斯本	Location：Lisbon, Portugal

埃斯托伊宫酒店是葡萄牙一家连锁奢华酒店。酒店历史悠久，不张扬的现代风格，让人流连忘返。标识设计中规中矩，颜色上选择褐色，与周围的环境和谐地融合在一起。

The Pousada Estoi Palace is a Portuguese chain of luxury hotels. The hotel is in a magnificent palace from the 18th century and combines unobtrusive way of history and modern styles. The no-nonsense typography fits harmoniously into brown-beige color scheme into its surroundings.

Signage

 希拉约翰逊设计中心
Sheila C. Johnson Design Center

项目档案	Project Facts
设计：Lyn Rice Architects	Design：Lyn Rice Architects
项目地点：美国，纽约	Location：New York, USA

设计中心的临街面有一个很特别的华盖，华盖是由一行标识的字母构成的。字母垂直立在门口，只有从正确的角度看，你才可以清楚地看到这些字母。在大楼的双侧，有红色的旗帜广告；在室内，捐赠者的名字非常有创意地设计在墙面上，而学校的名字则以孔洞的形式呈现，就像是礼堂的背景幕。

On each frontage a sign doubling as canopy marks an entry to the Design Center. What the canopy is actually "saying" isn't really clear until one looks at it from the right perspective. The red banners on the side of the building effectively convey to passers-by what is inside the building, so the canopy becomes a design gesture. Inside, donor names are creatively set into wall surfaces and the schools name is spelled out in perforations as an auditorium backdrop.

长崎县美术馆标识
Nagasaki Prefectural Art Museum Signage

项目档案　　　　　　　　Project Facts

设计：原研哉，色部义昭　　Design：Kenya Hara, Yoshiaki Irobe
项目地点：日本，长崎县　　Location：Nagasaki, Japan

美术馆入口的标识呈百叶窗形，看起来像是梳子的一根根锯齿，垂直地立在地面上，排成两行。当人们路过看到这个标识的时候，会意外地发现优雅的动态之美，独具三维的莫阿效应。

The design for the entrance signage to the Nagasaki Prefectural Art Museum uses architectural louvers like the teeth of a comb. Lined up in two rows, they stand vertically on the ground. When people look at them while walking past, there is a surprising dynamic movement, like a three-dimensional moire effect.

Signage

刈田综合病院标识
Katta Civic Polyclinic Signage

项目档案　　　　　　　Project Facts

设计：原研哉，小矶裕司　　Design：Kenya Hara, Yuji Koiso
项目地点：日本，长崎县　　Location：Nagasaki, Japan

刈田综合病院的标识可见度很高，字体大并且位于白色的地面上。为了防止褪色和毁坏，标识是由红漆排印的。室内各转角处都有清晰的标识，类似机场的标识那么明显，具有指导性。

The salient characteristics of the Katta Civic Polyclinic signage are high visibility due to the use of large type, and type laid directly into the floor. The designers came up with the solution of inlaying red linoleum type and symbols on the white floor to eliminate the problem of fading and erosion from foot traffic. As the complex had a flow line as clear as an airport's, the floor guidance graphics were clearly organized based on red crosses at the main intersections.

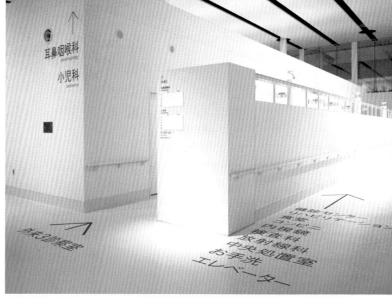

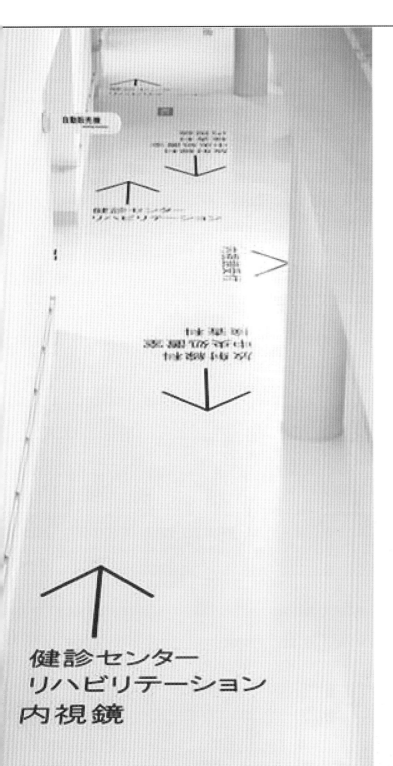

大型中庭图形化标识
Storehagen Atrium Graphic Signage

项目档案

设计：Ralston & bau
项目地点：挪威

Project Facts

Design：Ralston & bau
Location：Norway

本案标识设计注重简单性和清晰性，设计目的与地下的标识相一致。因为这个中庭将会变成一个重要的中心。鲜艳的颜色、生动的图像遍布每一个楼层，方便每一个人使用，即使是视觉上有缺陷的人。

The selected signage design is using the simplicity and clarity of the underground signage systems. The idea was influenced by the fact that storehagen atrium will be an important hub, a city with the desire to become a metropolis. Characteristic subway lines are used all through the signage system with strong colors and graphical shapes dedicated to each floor and institution, designed following the principles of universal design they should make it easy for any user, like persons with visual impairment, to find their way.

Signage

兰斯阿姆斯特朗基金会总部标识
Lance Armstrong Foundation Headquarters

项目档案	Project Facts
设计：Fd2s Company	Design：Fd2s Company
项目地点：美国，德克萨斯州，奥斯丁	Location：Austin, Texas, USA

Fd2s 设计公司将兰斯阿姆斯特朗基金会总部的公共空间转化为一个呈现总部使命、历史和成就的地方，同时也是为了感谢总部捐赠人。整个标识设计特点就是重复使用黄带，这也是总部最具有代表性和说服力的标识。因为总部是绿色建筑，所以在选材上都是采用剩余的物资。

Fd2s Company created a graphics program for the headquarters that turns the building's public spaces into a venue for conveying the mission, history, and achievements of the LAP and its many constituent groups, while also providing opportunities to recognize LAF donors. A recurring motif of the program is a yellow band with recessed or cut out type, which is a tribute of the foundation's most recognized symbol. The building is on track to achieve Gold LEED certification, and Fd2s contributed to this effort by working with the signage fabricator to use surplus materials from the fabricator's shop wherever possible.

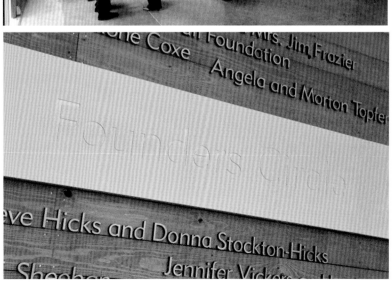

枫忒弗洛皇家修道院标识
Signage for Abbaye Royale Fontevraud

项目档案　　　　　Project Facts

设计：matali crasset　　Design：matali crasset
项目地点：法国　　　　Location：France

设计师设计了一个网络结构的标识系统，不管是有形的，还是无形的，目的就是突出各种道路，让参观者能够在这个古老的修道院里畅通行走。各种线型和横梁式的标识先分离后又汇合，具有动态感。标识实用性很强，而且具有情趣，可以清楚地告诉人们哪里可以行走，哪里可以参观，哪里可以休息。

The designer has designed the new signs as a network of lines, whether visible or invisible, that outline the various paths that visitors can take throughout the historic site. This concept of lines and beams separate and come together again, like extensions, information, directions which are practical and amusing within the context in which they are installed. These range of elements suggest staging the walk and visit, guiding the visitor to follow some directions more than others, or providing opportunities to stop, read or look for specific information.

Signage

医学院道路标识总体规划
Medical School Wayfinding Master Plan

项目档案

Project Facts

设计：Fd2s Company
项目地点：美国，阿肯色州

Design：Fd2s Company
Location：Arkansas, USA

本案包括创建一个导向标识系统，为新的住院部、癌症中心、精神病院设计新的标识。设计师同时也为Winthrop P. Rockefeller机构的捐赠者设计了一个综合标识方案，包括十二个室内和室外的标识设计，以此来感谢和纪念那些慷慨的捐赠者。

These implementation projects included the creation of a detailed sign standards manual, programming, documentation, and contract administration services for all signage elements in a new patient tower, cancer center, and renovated Psychiatric Research Institute. Fd2s also created an integrated program of donor recognition signage for the new Winthrop P. Rockefeller Cancer Institute, which is now being deployed at approximately one dozen different interior and exterior locations throughout the facility.

Signage

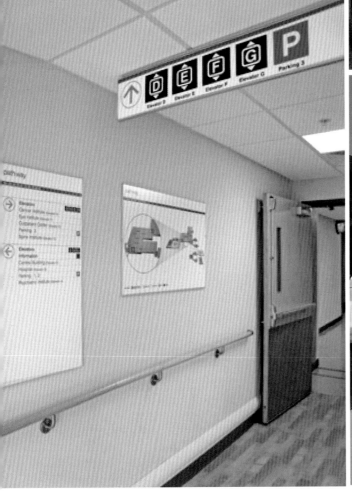

福尔斯克里特高山度假村
The Falls Creek Alpine Resort

项目档案	Project Facts
设计：Michel Verheem	Design：Michel Verheem
项目地点：澳大利亚，维多利亚州	Location：Victoria, Australia

福尔斯克里特的道路呈锯齿状往山上延伸，这意味着，如果你的目的地在西边，则必须往东边走才能到达。这完全颠覆了人们对地图的感知，因此一套能指示出景点的道路指示系统就显得非常必要。经过战略发展阶段，ID/Lab 开发出一套道路指示系统，这套系统的出发点在于什么信息需要在什么地方出现，并且以什么方式出现最好。系统包括了入口、建筑、小径开端、交通标志、方向指示、解释说明，还有零售、广告等标志。

Roads in Falls Creek zig-zag to climb up the mountain, which means that usually you have to go east to reach a destination west of you. This threw out people's reliance on their cognitive mapping ability, and showed that the system even, needed to direct to well known destinations. After developing the strategy, D/Lab created a wayfinding toolkit. This toolkit set out what information needed to be displayed, and how this information could best be displayed. It included gateway, building, trail-head, statutory, traffic, directional, interpretive, promotional, retail and advertising signage.

Signage

标识

ANZ 中心标识
ANZ Centre Signage

项目档案 | Project Facts

设计：Fabio Orgarato Design
项目地点：澳大利亚

Design：Fabio Orgarato Design
Location：Australia

Fabio Orgarato Design 公司与 HASSELL 和 Bovis Lend Lease 合作，设计出一套符合 ANZ 文化需求和工作环境价值的综合路线指示系统。
路线指示系统是建筑的一种延伸。优雅的雕塑造型灵感直接来源于建筑细节中的移动板。设计师还设计了一系列多样的超大图形和装置，以适应各种不同的环境色调，激发员工之间的互动。

In collaboration with HASSELL and Bovis Lend Lease, Fabio Ongarato Design devise a comprehensive wayfinding and environmental graphics system that matches ANZ cultural needs and workplace values.
The wayfinding system is created as an extension of the architecture. Elegant sculptural forms are directly inspired by the shifting planes that are detailed in the architecture. A diverse range of supergraphics and installations are created to suit varied hub environments that encourage staff to interact with each other.

Signage

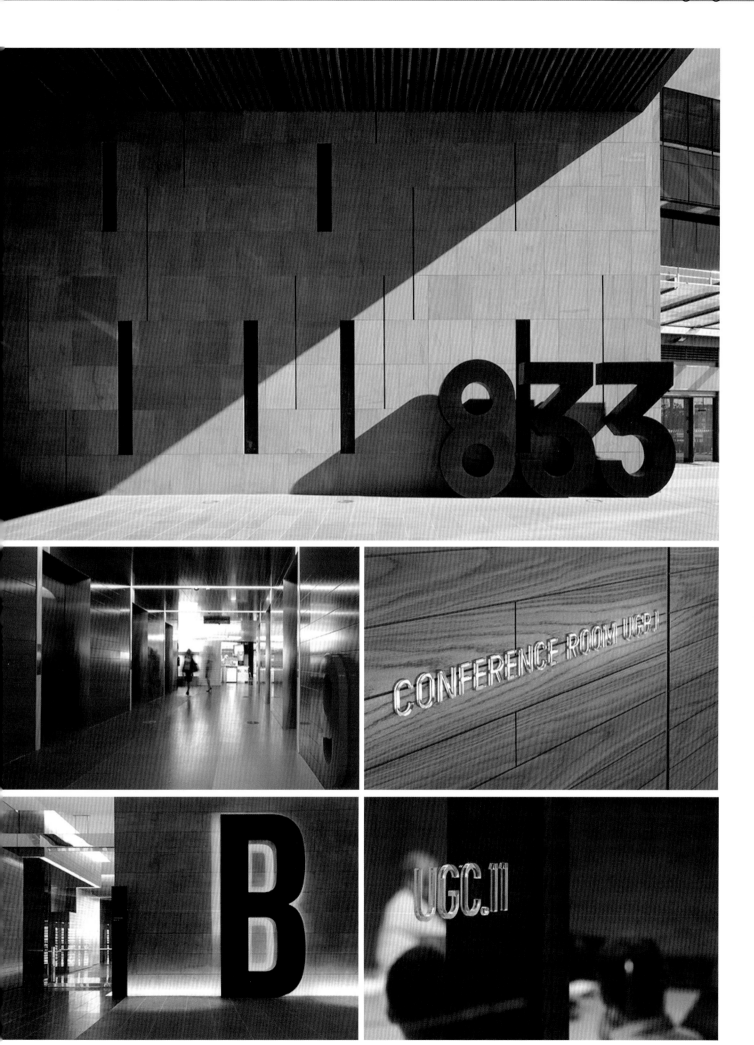

标识

Signage

Urban Attitude 礼品店标识
Urban Attitude Signage

项目档案

设计：Fabio Orgarato Design
项目地点：澳大利亚

Project Facts

Design：Fabio Orgarato Design
Location：Australia

因为新主人的到来以及向国际发展的新趋势，Urban Attitude 礼品店需要一个改变，促进品牌的重新建立以及扩大销售量。设计师们设计了一系列的标识，包括晶状体卡片、光纤、网络游戏、街景画等。

Urban Attitude is a novelty gift store. With new owners and an intention for retail expansion nationally, FOD were commissioned to undertake an extensive re-brand and retail experience for the future. FOD also developed a suite of collateral and signage, which included lenticular cards, optical patterns, online games, streetscape flags and illuminated signage.

成人教育中心标识
Centre for Adult Education Signage

项目档案 / **Project Facts**

设计：Fabio Orgarato Design
项目地点：澳大利亚

Design：Fabio Orgarato Design
Location：Australia

设计师们在本案的设计中选择了一种复杂的现代的方法，力求创造一种综合性的视觉语言，不仅信息量强大，且过渡自然还具有连贯性。在设计之前，设计师们先对整个空间作了一个本质性的了解。

The designers choose a sophisticated and contemporary approach expressed through a signage and environmental graphics that exploit a comprehensive visual language based on information highways, connectivity and transformation. The design project explores the notion of transforming lives through learning, recognizing the importance and changing nature of the organization.

Signage

标识

Signage

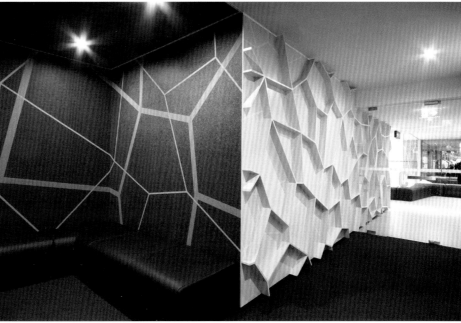

华盛顿大学萨弗里厅标识
Savery Hall, The University of Washington

项目档案　　　　　　　　Project Facts

设计：STUDIO SC 公司　　Design：STUDIO SC Company
项目地点：美国，华盛顿特区　Location：Washington DC, USA

华盛顿大学萨弗里厅导向标识系统综合性强，特点鲜明。标识上除了基本的导向信息之外，还有清晰的室内结构图，恰到好处地表现了建筑的工业化特点以及历史的丰富性。

The project is a comprehensive identity and wayfinding program that reflects the industrial character of the newly exposed interior structure and celebrates this unique buildings history.

Signage

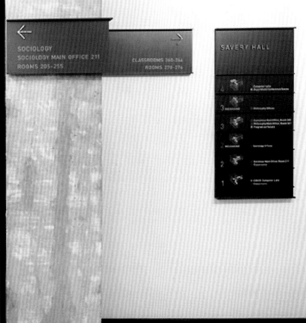

标识

费尔班克斯国际机场标识系统
Fairbanks International Airport Signage System

项目档案

设计：STUDIO SC 公司
项目地点：美国，阿拉斯加州，费尔班克斯市

Project Facts

Design：STUDIO SC Company
Location：Fairbanks, Alaska, USA

本案的标识是专门为费尔班克斯国际机场的整修和新的发展而设计的。标识都是简单的现代样式，反映了当地环境和建筑的特点。

The project is a comprehensive identity and signage system for the expansion and renovation of this international airport. Signage components are simple modern forms that reflect the character of local landscape and architecture.

Signage

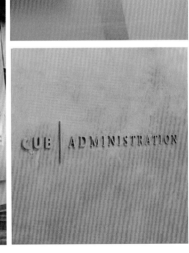

 ### 华盛顿大学康普顿工会大厦
WSU Compton Union B/D

项目档案	Project Facts
设计：STUDIO SC 公司	Design：STUDIO SC Company
项目地点：美国，华盛顿州，普尔曼市	Location：Pullman, WA, USA

本案将建筑的材料和形状融入到标识设计中，让人感觉这些标识是本来就存在于建筑中，平衡建筑环境和信息之间的关系。有了这些标识，人们可以高效且毫不费劲地通过各个空间。

By engaging building forms and materials with graphics, this comprehensive signage program illuminates wayfinding cues that are inherent in the architecture, yielding a balance between the built environment and information. This seamless design intuited leads users go through the space efficiently and effortlessly.

标识

 任天堂公司标识
Nintendo Signage

项目档案

设计：STUDIO SC 公司
项目地点：美国，华盛顿州，雷蒙德市

Project Facts

Design：STUDIO SC Company
Location：Redmond, WA, USA

本案的标识设计体现了任天堂企业总部干净现代的环境以及富有生命力的品牌形象。外部的标识体现了任天堂的产品以及 LOGO 特征。室内的标识设计主要体现在停车场，怪诞的设计让停车场显得个性、亲切。

This environmental graphics program reflects the exuberant spirit of Nintendos brands and the clean, modern environment of their new corperate headquarters. The design of the exterior signage and graphics reflect Nintendos product and logo. Graphics inspired by Nintendos Mario Kart infuse whimsy into the parking garage, making it personable and welcoming.

珠宝世界杂志社标识
Jewel World Signage

项目档案	Project Facts
设计：Arris Architects	Design：Arris Architects
项目地点：印度	Location：India

设计师们所面临的挑战很巨大，不仅要在有限的空间中最大化地促进销售，而且还要提供给客户较高的视觉满意度。本案设计的重点就是视觉上的交流，并最终创造了一些个性鲜明的店面，一个长115米并且逐渐弯曲的走廊由一组组曲线点缀着，店面的用材也各不相同。

Arris Architects were assigned an exigent task——not merely to maximize the retail footprint in a very tight structure but also to offer the patrons a visual gratification never experienced before.
Visual communication was the key design focus while conceiving the retail area, and the final design succeeds in exploiting individuality for each and every shop front. There is a 115-meter long and gradually curving corridor, interspersed by sinuous curvilinear elements, which "de-scale" the long corridor for the patrons, while bringing a sense of identity for each shop. The shop fronts on either side of the corridor are treated with distinctly different materials.

Signage

盖顶人行通道
Covered Pedestrian Crossing

项目档案	Project Facts
设计：Atelier 9.81 工作室	Design：Atelier 9.81
项目地点：法国，图尔昆	Location：Tourcoing, France
面积：150 平方米	Site Area：150 m²

图尔昆市中心正处于一个大规模的重建过程中，所有的公共空间、街道和广场都被翻修。带有影院的巨型购物广场也将在近期开幕，地铁、电车和公交车站将连为一体，为使用者提供一个真实的多形式的交通系统。这个新的盖顶人行通道也是在此项目背景下产生的，它为不同形式的交通方式创造联系。人行通道与一系列住宅完美融合，并呈现出相似的风格，但同时人行通道也定义了自己的视觉影响：橘红色的光泽应用于开阔的尖顶之上，另有透明的玻璃表皮在夜晚中被照亮。

走进这个空间，你会发现一个顶棚，这个顶棚代表了城市的手工折纸艺术。长为 20 米，宽为 4.5 米的闪银色铝塑板，展示了丰富的折痕以及高度。顶棚延伸到一斜坡处，通过这里可以到达地铁和购物中心，通道的地面是灰色的花岗岩铺设的，与整个城市的公共空间的地面铺设一致。

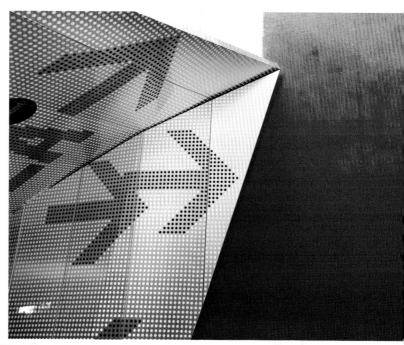

Signage

Downtown Tourcoing is currently at the heart of an extensive restructuring, launched a few years ago. All public spaces, streets and squares are being fully renovated, and a large shopping mall with movie theaters will be inaugurated soon. As part of this project, the metro, tram and bus station come together to offer a true multimodal system.

The project of a covered pedestrian crossing for downtown Tourcoing is born of this new direct relationship between transportation modes (with the bus station on one side and the trams and subway on the other), between the Place du Doctor Roux and the Place Charles et Albert Roussel. The pedestrian crossing will fit into a row of townhouses the same style, taking the place of one of them. By breaking thus with the alignment, the pedestrian crossing asserts itself visually, with the orange-red hues used on the open gables and by the constitution of a retro glassed facade, lit up at night.

Stepping into the void thus constituted, the project consists in erecting a canopy representing an urban origami. Spread out over a 20-meter-long and 4.5-meter-wide area, this sheet reveals numerous complex folds and height variations, from which it derives its uniqueness. The covering ends in a notable slope, signaling the pedestrian crossing from the tram terminus and the entrance to the shopping mall and metro. The crossing's floor is made of gray granite pavement, an extension of the planned layout for all downtown public spaces.

More Signage

Signage

标识

More Signage

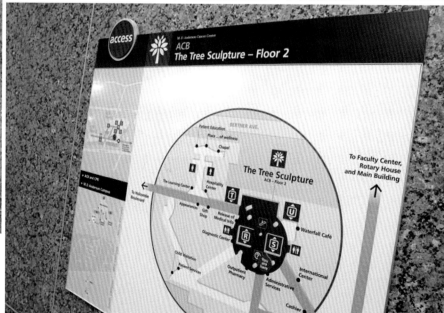

Signage

More Signage

Signage

标识

More Signage

More Signage

Signage

标识

More Signage

Signage

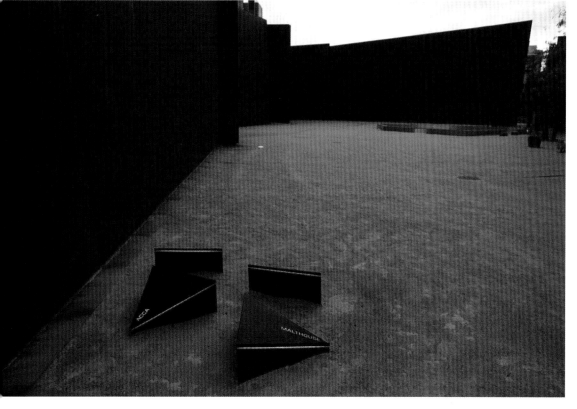

More Signage

Signage

More Signage

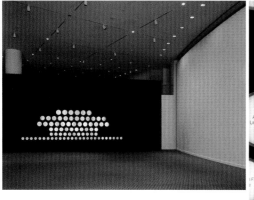

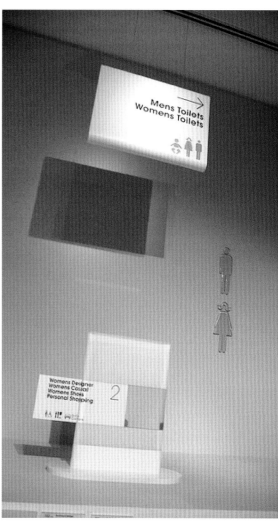

Signage

标识

More Signage

Signage

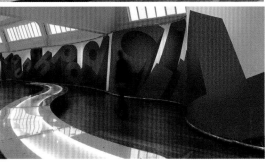

标识

More Signage

Signage

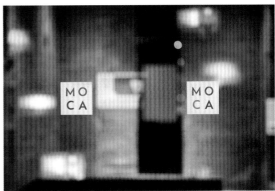

More Signage

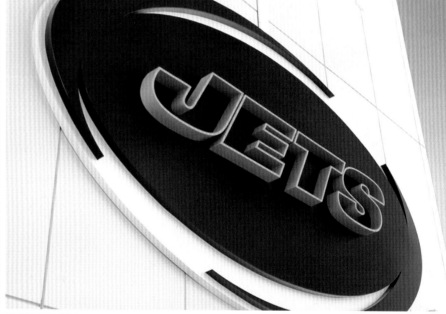

标识

More Signage

Signage

标识

More Signage

Signage

标识

Signage

标识

More Signage

Signage

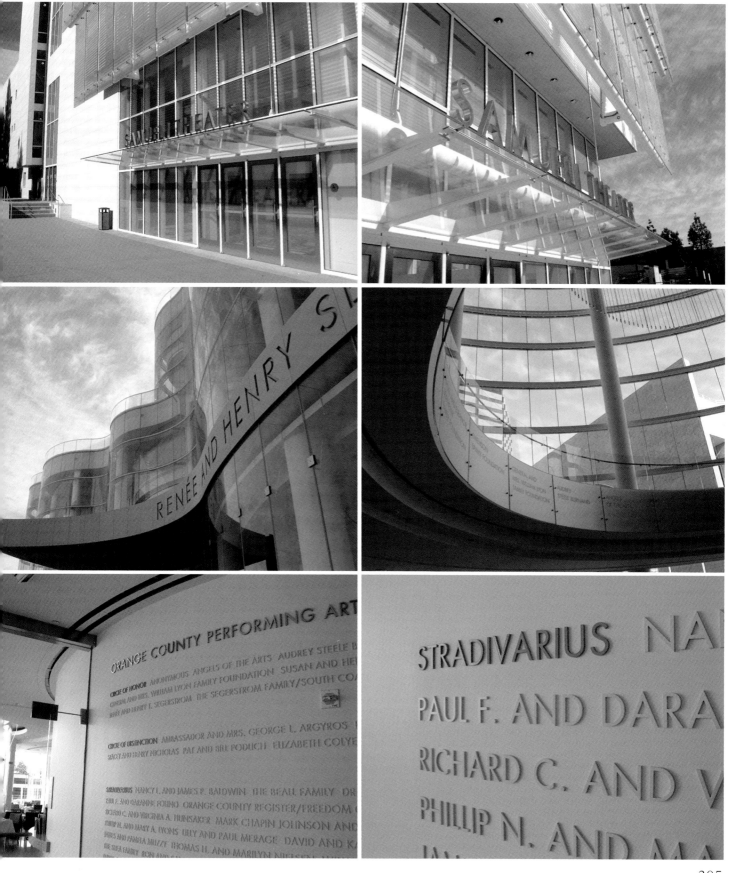

图书在版编目（CIP）数据

公共景观集成.建筑景观/广州市唐艺文化传播有限公司编著.－－北京：中国林业出版社，2016.4

ISBN 978-7-5038-8443-6

Ⅰ.①公… Ⅱ.①广… Ⅲ.①景观设计-图集 Ⅳ.① TU986.2-64

中国版本图书馆 CIP 数据核字 (2016) 第 050692 号

公共景观集成　建筑景观

编　　著	广州市唐艺文化传播有限公司
责任编辑	纪　亮　王思源
策划编辑	高雪梅
文字编辑	高雪梅
装帧设计	杨丽冰
出版发行	中国林业出版社
出版社地址	北京西城区德内大街刘海胡同 7 号，邮编：100009
出版社网址	http://lycb.forestry.gov.cn/
经　　销	全国新华书店
印　　刷	深圳市汇亿丰印刷科技有限公司
开　　本	220 mm × 300 mm
印　　张	19.125
版　　次	2016 年 8 月第 1 版
印　　次	2016 年 8 月第 1 次印刷
标准书号	ISBN 978-7-5038-8443-6
定　　价	306.00 元（精）

图书如有印装质量问题，可随时向印刷厂调换（电话：0755-82413509）